T0291274

Quality Beyond Borders

Perceptions as to the nature of the Quality Sciences and disciplines vary across the world depending on local industrial history. This can cause problems for global organisations who often want to retain the quality policies of the parent company whilst attempting to embrace the approaches familiar to local people. For example, whilst Western organisations have embraced Six Sigma, Lean and other Japanese management techniques, we have tended to adopt them in a hotchpotch fashion, bolting them on without ever understanding the context behind total quality control. In Japan, these concepts are not considered to be standalone but are all part of a seamless companywide matrix of interactive concepts, which can be summed up as company-wide quality work, of, by and for all. In essence, this means that 'quality' is everybody's responsibility from the chief executive downwards.

David Hutchins has over several decades worked in all of the cultural blocks and has consistently managed to integrate all of these differences into a single company-wide approach. When the concepts covered are integrated into a total company-wide programme, the intention is to make that organisation the best in its business; in Japanese terms this implies 'Dantotsu', which means 'number one thinking'.

Accessible and practical in approach, *Quality Beyond Borders* is split into short sections, each representing a self-contained idea for the reader to digest and reflect on. It is a valuable resource for business practitioners, students and academics alike that will enable you to reach beyond your own borders to implement new ideas with significant results.

David Hutchins has a master's degree in Quality and Reliability from Birmingham University, UK. He is a Chartered Mechanical and Electrical Engineer, Chartered Quality Professional, Fellow of the Chartered Quality Institute (CQI), a John Loxham lecturer and author of several books and many articles.

Quality Beyond Borders

Dantotsu or How to Achieve Best in Business

DAVID HUTCHINS

Routledge
Taylor & Francis Group

LONDON AND NEW YORK

First published 2019
by Routledge
2 Park Square, Milton Park, Abingdon, Oxon OX14 4RN

and by Routledge
52 Vanderbilt Avenue, New York, NY 10017

Routledge is an imprint of the Taylor & Francis Group, an informa business

British Library Cataloguing-in-Publication Data
A catalogue record for this book is available from the British Library

Library of Congress Cataloging-in-Publication Data
A catalog record has been requested for this book

ISBN: 978-1-138-56507-4 (hbk)
ISBN: 978-1-138-56510-4 (pbk)
ISBN: 978-1-315-12280-9 (ebk)

Typeset in Stone Serif
by Integra Software Services Pvt. Ltd.

Contents

Figures

Acknowledgments

This is going to read like an award ceremony at the Oscars but everyone that I am going to mention is very relevant, not just to this text but the inspiration that has driven me from my roots as a 15-year-old school leaver with no qualifications of any kind to this point in my life where, at aged almost 82, I have lost none of my passion and am working harder than ever with this being my ninth book and millions of miles on the clock in the aluminium tubes that I have slept in some 40,000 ft above the Earth's surface in order to teach, present, discuss and work with quality almost all over the globe.

I can attribute the start of my personal development to the IQ test I was required to take prior to my conscripted military service back in 1958. One month later, after mobilisation, I was asked to take it again. Following that, I was interviewed by an officer. He asked if I was curious as to why I had to take it twice. I said it had never occurred to me – I assumed it was Army procedure. He then questioned me on why I had been content to work in an iron foundry prior to that. Again, I shrugged my shoulders. With no qualifications what was I expected to do? It was a job. He then told me that I had an IQ of 142. They did not believe it the first time, so they did it again and got a similar score.

You would think I would have been happy, but I was furious. I thought about all of those wasted years when I could possibility have gone to university and had a good job with good prospects. I made up my mind that I was going to put this ability to the test, so I did. Three times a week I went to night school for eight years, at the end of which I got the equivalent of two degrees in Mechanical and Production Engineering.

Each year I passed I got more angry, then I had the opportunity to go to university to study for a master's degree in Quality and Reliability. It was when I got to university that I suddenly stopped being angry and realised that I was, in fact, privileged. Most people only live in one world, that of the successful or that of the what seemed to me to be rejects.

I then thought of all those millions of people who were in the world that I had left and that is where my passion was born. This just happened to coincide with my first contact with Professor Ishikawa, which derived from a meeting my brother-in-law Tony Shaw set up for me with the late Professor Sasaki at Ashridge Management College. Professor Ishikawa sent me some material that described Japanese TQM, which involved the entire payroll. This sparked me into an immediate realisation that this was something that could at a stroke change the whole lives of those I just described as being fellow rejects. The thought of involving the entire payroll in a concerted plan to be the best inspired me so much that it changed my life. So much so that I organised a conference at the Institute of Directors in 1979, with Professor Ishikawa co-chairing the entire three-day event and speaking for most of the first day. This resulted in an interview on World Service radio, my first book *The Japanese Approach to Product Quality*, then meeting up with Dr Juran with whom I worked for the following ten years. I developed a very close relationship with JUSE in Japan where I took many study tours and, in return, organised UK tours for Japanese organisations. I had many colleagues during that time. I will mention just one, my special friend and co-director, Brian Tilley, who sadly died last Christmas.

I was fortunate to become a close friend of Dr Kano who, although sadly I did not meet him before I met Dr Sasaki, I did arrange the publicity for the tour of the UK that he made in 1974. Those years up to 1992, when Dr Juran retired from international travel, were just incredible times. Other people who inspired me were Dr William Thoday who was quality manager at Rank Xerox Welwyn Garden City and for a year president of EOQ, the European Organisation for Quality. Bill was also one of the three original authors of the EOQ glossary of terms used in 'quality', which has since been adopted by ISO.

There was also Roy Knowles the secretary general of the Institute of Quality Assurance who never ceased to give me encouragement and later, many years after Roy's death, Frank Steer who was responsible for enabling the IQA to change its name to CQI and attain chartered status. Over those years apart from my business activities I was extremely active in the IQA and was for many years its representative as Chairman of the Human Factors Committee of EOQ.

About the time that Dr Juran retired, there was another boost to my passion for 'quality'. After speaking at a conference on quality circles in Hong Kong, I was followed onto the platform by four school children from the CMS School in Lucknow, India. They we aged just ten. They made a quality circles presentation that totally overwhelmed both me and the other 1500 delegates.

We learned that the school was in the *Guinness Book of Record* with over 28,000 students from kindergarten upwards. All students over the age of ten were encouraged to form quality circles. The head of the school, Dr Vineeta Kamran, met me afterwards and she told me that the principal, Dr Jagdish Gandhi, had, two years before, visited Japan, and did an industrial study tour. He discovered Japanese TQM and quality circles and formed the opinion that it was a good idea to experiment with it in education. I, together with another keynote speaker, the late Don Dewar, suggested that the school should organise a conference for the world to see. In 1997, the school followed this advice and staged a conference. It was phenomenally successful. So much so that it was repeated in 1999. Over 3000 people and students attended from India and surrounding countries. From there it has been repeated every year with each alternate event taking place at Lucknow. This year it was in Bangladesh and I have attended and been a keynote speaker at all except three at Lucknow and many in the other countries, which included Kingston University in the UK where the orchestra of our local primary school played at both the opening and closing of the first day. In Nepal alone, students' quality circles are now part of the National Curriculum with an estimated over 28,000 students involved. I am proud to know both

Dr Gandhi, Dr Kamran from the CMS School and also Dinesh Chapagain who is credited with starting the movement in Nepal. All of these and others graciously tell me that I have inspired them, but I can say quite categorically they have unquestionably inspired me, and I am blessed to have them amongst my friends. Without them, I am sure this book would never have been written.

Now back to home, my family and friends. Friends first. Over the past 20 years my closest associate has been John Dansey. Not only has he been as solid as a rock through good times and some not so good, John has been there. Not only that. John loves systems far more than I do. Anything related to that aspect of the subject John deals with, which frees up my time to do other things. You are a blessing, John! Now, finally, family. Where would I be without my wonderful wife, Margaret. I am not getting all slushy, but seriously. We met just after I was demobbed from the Army and I was obsessed with my drive to recover from my lost education. There were loads of guys who would have been all too willing to entertain her whilst I was slaving away at night school, but she stayed with me. For the first few years of our married life I was still slaving away and all I got was encouragement and it has been like that ever since. She is still working with me. She gets a bit cross at times but we are still here!

Then there are our daughter and son, Caroline and Michael. In recent years, Caroline has joined the business and has had a remarkable effect. Caroline has skills that go way beyond mine in being able to take my rough and ready ideas and make them into professional-looking products that will equal anybody's. Mike is a different person altogether. Way back he came with me on a few trips including work in Tunisia and Iran. I had hoped that he would join me, but he has done much better than that. Marrying his lovely American wife, Erin, the pair started a business in Washington DC, which in no time became a world beater in the field of coaching. Being ICF certified, they rank amongst the best in the world. Of special interest to us, many of their students from large international companies are quality engineers and, as a consequence, DHIQC has adapted its top-level units to make a seamless join into their field as can be seen by the competency wheel that Mike created for DHI and that is included in the final chapter of this book.

It is all the above that has inspired me. There are literally hundreds of people that I would have loved to have been able to mention but I could go on for ever!

About the author

David Hutchins has a master's degree in Quality and Reliability from Birmingham University UK. He is a Chartered Mechanical and Electrical Engineer, Chartered Quality Professional, Fellow of the Chartered Quality Institute (CQI), A John Loxham Lecturer and author of several books and many articles. In David's early career he was Chief Production/Industrial Engineer in the Automotive Components Industry before becoming Works Manager followed by 10 years teaching and consulting in Business Management prior to founding David Hutchins International and its International Quality College. David Hutchins has over forty years of continuous experience in all aspects of the Quality-related sciences on a world-wide basis. He co-presented with the unchallenged World leading expert, Dr Juran, who died in 2008, on all his annual courses in the UK from 1983 until Dr Juran's retirement from international travel in 1992. He was a personal friend of the late Professor Ishikawa and David was the only European to be invited to contribute material for the book which commemorated his life. He has been a key note speaker at conferences all over the world including many seminars organised by the Union of Japanese Scientists and Engineers (JUSE) in Japan and to this day works with the well-known specialist in Concept Engineering, Drs Kano and Shiba.

Tips on how to use this book

Whilst writing the book, it was never my intention that it would be read from cover to cover like a novel. If it is, then I estimate that probably some 80% or more will be quickly forgotten if absorbed at all.

It was written based on the observation that when presented with some possibly new ideas, the brain needs time to assimilate them, question them and then to match them against possibly different or opposing views before either accepting or rejecting them.

Also, and certainly more important, is the time required to think about how the idea might be used to advantage not just in work but in many cases, one's domestic live, pastimes and sports. This is important because there is a lot of philosophy contained in this book and it is as much a way of thinking as it is some mechanistic concepts that can just be bolted on to what is already there.

In fact, I can assure you that between its covers you will find ideas that might help you catch that biggest fish, win the 100 metres when previously you always came second, improve the overall quality of life and to help you turn your organisation into a world beater.

To help you to do all of that, the book is split into nine sections, each apart from the first rather general chapter, covering a specific aspect of the subject. However, each of these chapters are broken up into approximately ten short sections, each of which represents a self-contained idea that needs some thought and reflection.

First of all, buy a note pad specifically for use with this book. It is recommended that the reader takes these short sections, one by one, does not read any more that day and consciously thinks about it in terms of its relevance to the workplace, home and other interests. For each section, make notes on your thoughts related to that content.

Only when you have thought this through, maybe discussed it with family or colleagues should you move on to the next section. There will be some sections that maybe you will want to think about a lot. There will be others that do not seem particularly relevant or that you are doing them very well in your opinion any way. Well even there, do not be too impatient, maybe you have overlooked something.

If you use the book this way, it could take weeks or months to get from cover to cover. Well, it is not a race and if you have got something out of it, it will be worth it.

Now back to your note book. If you have studiously been taking notes and making observations, you should quite easily be able to convert that into some sort of action plan to move those things that are under your control in the direction that you want to go. Maybe you have shared this with colleagues. If so, and you think all of this was worthwhile, then I will be a very happy author.

Preface

POST-WORLD WAR II QUALITY HISTORY

The Japanese quality revolution

Initially, I had not intended to begin the book by mentioning this topic but the more I thought about it the more I realised that I had little choice. This history is just so relevant to everything related to quality, to business performance and the state of Western economies that if I did not begin here, the remainder of the book would be hard to follow. Also, it is my opinion that there is so much misinformation and fake news about this aspect of management out there in the market place that it is a serious block on being able to understand the big picture. If you do not understand that then I fear that you will fail because you might be too easily influenced by those who I would say have been radicalised by strategies that look good on the surface, but that have little in the way of genuine substance. It is a case of both 'buyer beware' and 'all that glitters, is not necessarily gold'.

I am mindful of a quote attributed to Winston Churchill 'history is written by victors'. I think this is very much the case here.

As mentioned in the Introduction, some key aspects of post-World War II Japanese history have been included in Appendix for those who are interested.

Let us briefly look at the relevant history in a bit more detail.

The USA

The big changes began in the late 1800s. The so-called 'division of labour' method had existed right back to the days of Socrates and Plato. Even Adam Smith commented on it in one edition of *Wealth of Nations*. Also, Karl Marx famously said, 'under this method [capitalism], the arms of the bosses will get fatter whilst the arms of the workers gets thinner'. It is reasonable to begin with a brief discussion of the social Industrial Revolution brought about by Frederick W. Taylor, Frank and Lillian Gilbert, etc.

In the USA these and several of their compatriots developed what they referred to as scientific management. The concept was as cold as it sounds with the work being systematically deskilled and reduced to the most mindless of operations. For example, in Gilbeth's 'therbligs', which is Gilbreth backwards, work is even reduced to the blink of an eye (referred to as 'blinks'). But there were huge productivity benefits derived from this dehumanised approach as illustrated by Henry Ford in the production of the famous Model T in which the sales price for a saloon car plummeted from around $300 per vehicle to just $59!

Not surprisingly, it was not long before this concept spread across the USA, raising productivity everywhere and resulting in the USA becoming the most productive country on Earth.

Interestingly, although as mentioned in the Introduction, the deskilling of work began with the introduction of unskilled female labour in military production during World War I, it did not spread to UK manufacturing generally, until after World War II.

Prior to that, almost all products in the UK were made using the principles of craftsmanship. A craftsman is responsible for the quality of his or her own work. It invokes a sense of pride. The foreman or supervisor was usually the most skilled person, highly respected by both management who depended upon him/her and the workforce who looked to him for guidance and generally this is how UK manufacturing did things until well into the 1950s.

Also, whereas a craftsman could and would inspect his/her own work whilst performing operations, because of the lack of training and personal development, there grew up in the UK a new work culture where inspectors were enlisted to catch any defective work before it reached the customer. Sadly, all of that was destroyed (one could say in the 'blink of an eye'!) in the late 1950s when Harold Macmillan was running an election with the popular slogan 'you never had it so good'. Well, this might have been true for material wealth but there was an unwelcome social consequence. People resented being treated like robots.

The full development of the division of labour with all the relevant skills of method study impacted the UK with the same social consequences as in Japan. However, the Japanese reacted to this very differently from the UK and so the main diversion in approaches to quality management began.

It was around this period that the industrial culture of the USA and that of the UK and the rest of Europe began to merge. American management consultants who could even claim their train slowed down going through Harvard could claim astronomical fees in the UK.

The main difference, though, between the UK and probably most other countries was that whilst the initiative for this approach to management came from high volume commercial manufacturing outside of the UK, in the UK, the initiative was squarely with the Ministry of Defence and its sophisticated procurement approach. For example, in the late 1960s it was claimed that the MOD was responsible for some 16% of all manufacture. This made them extremely influential. Because their approach was predominantly inspection orientated, they were not so concerned as to whether defects were made in the first place. This was not their problem, even though it must have reflected in the selling price. Rather, that the defects could be confidently segregated out before reaching their military destination. This was achieved by forcing suppliers either to set up sophisticated inspection systems and using prescribed sampling methods or by allowing an MOD inspector to reside in-house and administer this himself for the duration of production.

In the USA, on the other hand, they were more concerned with avoiding making defects in the first place! The consequence was that whilst Americans listened very carefully to the likes of Dr Juran the famous quality guru and others including Armand Feigenbaum, Phil Crosby and a host of others, most of these specialists were largely unknown in the UK right through until at least the 1970s, but still the processes were fundamentally based on the management systems created by Taylor and Gilbreth et al. and practised by Henry Ford and other powerful industrialists. They were convinced that this was the only way to manage and many industrialists operate this way overtly even to this day. Instinctively, to them it is the 'right way to manage'.

JAPAN

It is now 1945. Japan has just been defeated in the war in a dramatic and devastating fashion. Not only were two major cities, along with their populations, industries and interconnecting infrastructure, obliterated by the atomic bombs but this damage, horrifying though it was, was small compared with the overall effect of massive incendiary raids on other Japanese cities such as Tokyo and Yokohama, etc. The population was emaciated, devastating malnutrition was rife and from 1945 until 1951 Japan was an occupied country.

They were not allowed to build ships, planes or to make steel during those postwar years.

Not surprisingly, the occupiers were arrogant. It never crossed their minds that Japan could go from that situation to becoming the second largest global power in less than a decade from when they finally left.

For example, from 1952 when they were first allowed to build ships again and produce steel, they went from being ranked zero to replacing the UK as the world's largest ship builder before 1962, in just ten years. Soon after that, they could boast of having 17 out of 23 of the world's largest blast furnaces. By halfway through the 1960s they had wiped out the UK motorcycle industry and so it went on, industry after industry, white goods, brown goods, automobiles, almost everything.

How did they do that? Well, here is where real history and the fantasies of Western accounts of that history, violently clash. The real history is plausible. What many Westerners believe is not.

Here is the Western story. Fact. General MacArthur was the supreme Commander of American Forces in Japan at the time. MacArthur's strategy was clear.

1. To help Japan to recover so that it would not have to live on handouts from the USA for ever more (some say the Americans might have overdone that part of their policy!)
2. To change the industrial culture so that it was unlikely ever to wage war against the USA again
3. To introduce Americans scientific management (the so-called Taylor system) where an almost feudal system was the current norm.

To achieve (1) above, in collaboration with the Japanese, a scrap and build policy was introduced. Companies could apply for this US aid but 30% of it was to be used to destroy an old plant and machinery and replace it with new.

With regards (2), there existed in Japan an organisation of top industrialists called the *zaibatsu*, and this dominated Japanese work culture and methods. It was heavily influenced by Panasonic and Mitsubishi as can be found in the Appendix. MacArthur banned this organisation and replaced the top people with young managers who had not yet become indoctrinated into the old style of management, so these young people were forced into higher management with nobody to copy so they had little choice but to copy what the Americans taught them.

MacArthur made this easy by setting up a training programme for upper managers under the SCAP agreement. The person appointed to manage this education and training was Homer Sarasohn and he was assisted by Charles Protzman who was a very experienced senior manager from Bell Labs, which, incidentally, was where Dr Shewhart the originator of statistical quality control (later renamed statistical process control) was a senior executive and where Dr Juran had worked up to World War II and in which Dr Edwards Deming conducted his PhD thesis.

The three-week courses on production-related concepts originating in the USA were conducted by Sarasohn, Protzman and another, Dr Polkinghorn, ran more or less from 1945 to the Peace Treaty in 1952 but it is reported that the Japanese themselves ran them for another 23 years after that. During those courses, Sarasohn brought invited subject specialists in from the USA to give keynote talks on specific subjects. Shewhart was invited but due to ill health was unable to accept the invitation and so Dr Deming went in his place.

Due possibly to the embarrassment of Japan having been able to recover as it did, the Americans were desperate to find an American hero who was responsible for Japan's recovery. Unfortunately, some self-seeking American consultants latched on to the fact that the Japanese had named their national quality award, which had pre-existed Deming's visits, but did not have a title, the 'Deming Award' in recognition for his work. Later, the Japanese offered a similar accolade to Dr Juran for their 'all Japan' quality award, which can only be achieved five years after winning the Deming Award. Accounts of that offer say that when offered, Juran, possibly due to humility, hesitated, which the Japanese took to be a negative. There is no such word as 'no' in the Japanese language.

In fact, the Deming story has now become so entrenched in the West that, regardless of the facts, regardless of the wealth of contrary data freely available on the internet, people will believe what they choose to believe. Fine, that is their prerogative but, sadly, for as long as the real history is smothered, false choices of quality-based strategies will be made. Therefore, there is little chance of them ever being able to compete successfully with those countries that are more enlightened, know the real history and follow what has been proven to be sound logical thinking and who have developed their organisations accordingly. I am referring not just to Japan but also China, India and a host of smaller countries in the Pacific Rim area. If this book makes even a small contribution to that rethinking, then the slog of writing it will have been worth the effort. Enough of history now, let us get down to the important things. Chapter 1 skims the surface of some of the key considerations when deciding on a choice of future business strategy.

As mentioned, I have included some of my research into that history, which can be readily downloaded from the internet for those who are interested, as an appendix to the book.

Introduction

The first thoughts that come to mind when deciding to write a book is 'who are my intended audience'? and 'what is it that I feel compelled to attempt to convey'? The converse is also true. 'Who am I not writing it for'? Well, every author is different and would possibly come to a different conclusion to these questions depending upon motive, but dear reader, I will attempt to answer them from my perspective. First, in my mind whilst writing, I am mindful of all those millions of working people around the globe who probably would not pick up a book on the subject of 'quality' because either they did not think it has anything to do with them, or they might think 'we have a quality manager, I will mention it to him/her, it is not my problem'. Second, there is a high likelihood that some of my opinions, depending where in the world they live, will not align with those of many in the local so-called quality profession because 'this is not the way we do things around here!' or that I have either left out or that I have ignored or misrepresented some aspect of the subject that for them is the centre of the universe. Well, I will answer that now. First, please reread point number one. I am not primarily writing this for the 'quality specialist' as such, but I would be delighted if they do read it and think carefully about the content. If I have left out or appear to have misrepresented some aspect of the broad subject, it could, of course, reflect my ignorance, but hopefully it is more likely that I have done so deliberately. We all have different perspectives on everything. I acknowledge that I am quite opinionated, and I am using this opportunity to share my opinions. They are arguable as everything is. I hope they are thought provoking because that is part of my intention and I am not afraid to be proved wrong. There is no point in reading a book if you already know all its content. All I will say is that with more than 50 years of coal-face involvement in the subject, I have not formed my views lightly and I would not state them if I did not think I had very good reason for thinking them or feeling passionate about the need to share them. I know very well from much of the material on social media that I am at variance with some who articulate their views, but variety is the spice of life it is said!

What I hope more than anything is that for those who until now have, for whatever reason, had a very narrow view on what exactly is 'quality', that it will open their eyes to realise that they have spent their lives so far missing out on what is the most exciting management philosophy ever to have been created. Believe me, to see 'quality' as I see it, then, for many, will be a life-changing experience. Negatively, I do not have the confidence to believe that I am so good an author as to be able to help you make that transformation, probably you must make it yourself and you will need to work on it. This book might not actually be the cake but I very much hope that it has most of the necessary ingredients. Maybe also you need to be a good cook!

QUALITY BEYOND BORDERS? WHY THIS TITLE?

This will be the ninth book that I have written over the past three decades and choosing a suitable title is often almost as difficult as writing the book itself. It is like choosing a name for your first child.

I have known what I have wanted to write for several years now and during that time I have considered an almost endless number of alternatives. Some two years ago I thought that I had finally settled on 'Quality without Borders', only to discover some time later that a longstanding friend of mine, Lennart Sandholm, had used that as the title of his autobiography and I guess that I had subliminally absorbed it and thought I had invented it!

So, the label I have finally settled on is only a small variation on that, but I think, fortunately, it more accurately describes what I have in mind.

Back to the question. The title is the result of five decades of intense involvement in the quality sciences and disciplines, not just in the UK but significantly the Far East, South East Asia, Continental Europe, both North and South Africa, the Middle East and the USA. During that time, I consider myself to have been very lucky indeed not only to gain an almost intimate experience of the different cultures but also the opportunity to study first-hand the differences in approach to the application of the Quality Sciences and disciplines.

My impression obtained from all of that experience is that there are three main and distinctive blocks. Their histories can be traced back to approximately WW1. They have some similarities with each other as there has in every case been quite a lot of cross fertilisation, but also there are some striking differences. These blocks are the USA, Western Europe and Japan. Then there are other countries or blocks whose quality-related development started much later and in most cases are the result of cherry picking what they perceived to be attractive in the spectrums of each of the dominant three.

The tendency has been in that respect to be attracted to what the more local source has been doing so there is a degree of cultural spread, which is to be expected.

To sum up the differences in as few words as possible at this point, both the USA and Western approaches have been heavily influenced by the fact that the evolution of the Quality Sciences have their roots in the offline inspection of high-volume production. In the USA it developed because of being the first country to introduce mass or flow line production based on what is referred to as the Taylor system (Fredrick W. Taylor) and slightly predates WW1. In Europe, the UK is often credited with the introduction of the inspection system because of the huge loss of life of its skilled, mostly male, craftsmen and their being replaced by unskilled women who in that day and age were previously housewives. To use this resource, it was necessary to a create management system not dissimilar to that which had developed in the USA. The main difference was that in the USA the skill was taken out of the work and therefore the product was inspected by others. This approach was used across industry, both military and domestic. In the UK, however, it was almost exclusively related to military production and the Inspectors were employees of the military procurement bodies and not employees of the supplier companies. The most common of these inspection regimes was known as the 6/48 and 6/49 terms of contract. As military procurement was responsible for the purchase of some 16% of all manufactured product, the policies adopted by this authority had a major influence on the future development of quality in the UK and, later, Europe generally.

In Japan, it was different again. Whilst there had been some migration of methods from the USA to Japan before World War II, this was mainly the absorption of specific methods and techniques such as some of the basics of statistical quality control. The fundamental management style had remained constant right through to the end of the war and the start of the American occupation. It was during that period that dramatic changes occurred. This detail of history is hotly disputed, but I have included an appendix where some of the evidence can be seen, which hopefully will convince some of the real situation. The dramatic development, regardless of the details of who did what, was partly due to the Americans virtually forcing their managerial approach onto Japanese industry. At least as important was the Japanese reaction to this. They did not adopt American methods as prescribed, but not only did they adopt and adapt but they were smart enough to figure out concepts of their own. The result was the creation of a unique blend of concepts, some absorbed from the USA and some of their own creation, The result was a unique managerial approach that focussed everything upon harnessing the collective skill and creativity of the workforce to make the implementing organisation the best in its particular field. This has proved devastatingly effective as history shows and resulted in the extraordinarily powerful concept of *Hoshin Kanri* briefly described in this book, which is the subject of my previous book of that name.

Following all of this, as other countries have emerged such as Brazil, China and India, these have studied these methods and, depending upon what they have found to be attractive, have absorbed and, in some cases, adapted what has impressed them the most.

Interestingly, it is the author's view that what has consistently produced positive and impressive results regardless of local culture are adaptations of mainly what has been learned from Japan. In the view of the author, there has been a cherry picking of selected bits of that philosophy by commercially minded Westerners with largely self-seeking motives. The success of doing it properly, is, without doubt, the case in both China and India, but it can be seen in degrees elsewhere. The converse is also true. Whilst there has been widespread implementation of concepts derived from the inspection-based approach in both the US and UK cultures, there is no evidence that this has had any major positive impact on business results. Unfortunately, the consequences of this history have become so deep rooted as to be extremely difficult to confront regardless of there being, very obviously, some far more convincing alternatives.

Beginning with Professor Ishikawa in 1974, I gradually made the acquaintance of several of the epoch-making gurus in the subject, both Japanese and American. In some cases, I cannot claim to more than to have listened to them making presentations but in many cases not only did I know them personally, but I worked with them as well. I refer to both Professor Ishikawa and Dr J.M. Juran. I have always believed that I must have been born lucky not only to have known them, but also a wide spectrum of their colleagues and associates whose profound knowledge has largely shaped my life and, whether still living or dead, will continue to influence my thinking until my time is no more.

Such was my relationship with both Dr Juran and Professor Ishikawa that I was an invited keynote speaker at both of their 100th anniversary celebrations and a chapter written by me is included in a limited edition of the story of Dr Ishikawa's life, which is placed on a shrine in Japan to his memory.

Through my career, I have been fortunate enough to have worked in and implemented quality-based change programmes in over 40 different countries. The most important thing that I learned for all of this is that whilst perceptions about the sciences of quality differ markedly from one part of the world to another, in fact, all the variants, both the best and the least useful, can work in any socio-political system without the need for any form of adaption. As I once heard a speaker comment 'quality concepts have no passport'! The fact that there are differences, owes more to historical reasons and to the most prevalent information paths.

For example, the UK, in general, tends to follow the lead of the USA in the use of management concepts and this has been the case since World War II.

For almost two decades after World War II there was very little change in the European approach until the early 1970s when, almost out of the blue, the UK Ministry of Defence made a huge investment in the promotion of what had previously been a largely unknown set of NATO documents known as the Allied Quality Assurance Publications (AQAPs). With some minor tweaking of the content, these were renamed 'The Defence Standards' and imposed on British military products suppliers in 1974.

Later, the content of these Standards became adopted and renamed British Standard 5750 and in 1987 resulted in the publication of an update called ISO 9000. Initially, this impacted Europe more than elsewhere and for several years it was resisted in Japan and much of the rest of the Far East.

However, soon after that, Japan became concerned about growing hostility in Western countries and particularly France to the ever-widening trade gap. This resulted in some clearly protectionist strategies involving the terms and conditions that countries adopted to make excuses to reject Japanese imports. One of the reasons given was that despite their considerable superiority in quality, the Japanese products were not ISO 9000 certified. So, reluctantly Japan decided that for its own protection that it would adopt ISO 9000 and my organisation had a significant role in the early training in Japan. However, this was not then, nor has been at any time since, at the core of Japanese quality strategies.

In those early years after the war, Japan had developed a unique approach to the application of the quality sciences and disciplines. These were enshrined in an award process known as the Deming Award and to this day the award is the backbone of the quality movement in Japan.

However, this was not only the case in Japan. The rise of Japan as a major industrial nation in the years following the war was so dramatic that its methods were very soon adopted by the surrounding nations and especially China and India. In fact, all the Asian side of the Pacific Rim have quality management approaches that are more biased towards Japanese philosophy rather than Western. It is true that almost all of them use ISO 9000, but whereas in the West the quality industry is almost totally focussed on the application of ISO 9000 in its organisations, in the East, it is there, but by no means a significant feature in any organisational strategy. For example, at the International Quality Congress held in Tokyo some two years ago, there were approximately 240 papers from delegates from over 35 countries and not one of them mentioned ISO 9000!

Having spent almost 50% of my working life in all these cultures, it has been easy for me to see the relative merits of the various approaches and the variants. In fact, this is something of a simplification of reality because there are not simply two distinct approaches, as such. I have been generalising so far to make a point. There also many variants in the USA that are strongly based on the work of one or more of the American post-war gurus who are regarded as being almost saintlike by not only Americans but others as well!

Taking all of this into account, what seemed obvious to me was the fact that there was no either/or. None of the concepts that were more prevalent in one society than another, were culturally based. Rather, they were the result of historical development. Even though we now have global communications on an unprecedented scale, interestingly there remains a huge level of ignorance as to management practices in one part of the world or the other. Furthermore, there appears to be very little interest in studying those differences and taking account of the relative merits. The concepts included in this book are those that, from my experience, have impressed me the most. It is not a catalogue of every concept that I can think of but those that I have on numerous occasions implemented in a spectrum of cultures and that I have seen demonstrated to me that really work. The book is written with the intention of showing that the careful use of an amalgam of all these approaches will produce a result greater than the sum of the constituent parts. It is as close as I can get to be a 'how to' given that not everything is relevant everywhere. A hairdressing salon is very different from a major steel plant! But I have tried to highlight what I think to be points that need to be thought about by any reader who plans to use the book to improve their own organisation's performance. You will find some relevant and others maybe not. Hopefully, by the time you reach the final page, you will have sufficient notes to enable you, and others, to create a plan for the systematic improvement of your own organisation.

My expectation is that if you use the book this way, you will find, as the title suggests, that you will have reached out beyond your own borders but hopefully have found some ideas that are new to you and which, if you implement them, do produce significant results.

The book is based on the premise that what every organisation wants is a managerial approach that makes them the best in their business! I hope this book will help you get there.

Incidentally, the text is also the base material for the DHI Quality Fundamentals books and the Diploma in Quality Leadership. If you or your colleagues enrol on any of these courses, you will also have the opportunity for one-to-one assistance.

Finally, this book is also complementary to my last book *Hoshin Kanri – the Strategic Approach to Continuous Improvement.*

Good luck!

1 *What is quality?*

First, a short note on the style of the book: from here on, as mentioned in 'Tips for using this book', from time to time you will find a prompt for you to think about what you have read and attempt to relate it to your work. You do not have to do this but if you do, then collectively your comments and observations will act as an aide-memoire for you to, if you wish, upgrade your organisation's current management system. You do not have to do it. If you are a DHI student, you might do it to help with your course assignments as this book can be used as a companion to the online course materials.

AIM OF THIS CHAPTER

The learning outcomes of this chapter are:

- Understanding how 'quality' and its features can improve our lives both at work and at home
- Recognising the impact of 'poor quality' on both us and our organisation
- How to identify the likely biggest costs of poor quality
- Understanding our personal role in quality improvement
- Understanding the importance of each of the first five 'common to all personnel' chapters in this book
- Understand how the quality function can help us to work more effectively
- Understand the eight management principles on which Western approaches to quality management is founded
- To understand how the so-called PDCA cycle of quality improvement is or should be present in everything we do and what can go wrong if it is not
- To understand how the principles of quality apply right through all of our processes.

SECTION 1: PERCEPTIONS

The probability is that if you are resident in the Western hemisphere it is highly likely that your perception is that 'quality' is the responsibility of the 'quality manager' or if not then someone who has the management of quality in their job description. Quite possibly, you will also be forgiven for thinking that quality is neatly wrapped up in a Standard named ISO 9000. Regardless of where you happen to be, unless you happen to travel a lot you may be forgiven for assuming that approaches to quality management are the same the world over. If this is the case then, whilst we understand why you might think it, in fact you couldn't be more wrong. Why?

This is actually a good question and one that even quality 'experts' debate all the time. The fact is that it is contextual. In other words, the meaning changes slightly according to the situation. It would be quite reasonable to say, 'I cannot define quality' but I know it when I see it. It is the same with all judgmental terms such as 'nice', 'beautiful', 'excellent', 'love', 'noisy', etc. Quality really is about being perceived to be good at something. We all want to be good at something whether it is our work, a pastime, in our community or family. If we have children, we probably want to be regarded by our children and friends as being a good mum or dad. As a family, we want to be respected and have our opinions requested and taken seriously.

If we participate in team sport, we want to be selected; we want to play well. We want to get on well with our fellow players and collectively we all want to win.

If it is an individual sport such as cycle time trialling or track athletics, we might still want to be selected and be part of a team but our build-up and training is often down to us or us and, possibly, our coach. You might not realise it but in doing so you are actually practising what I regard as being *quality principles*. So, without realising it you are on the way to be a 'quality expert' anyway. Let us try to refine all of that!

'Quality' is all of the above, trying to do our best, knowing what is regarded to be 'good', knowing our own performance, having the means to check our own performance against some standards and then making improvements where we see the opportunity.

So, 'quality' is just as important outside of work in our private lives as it is in our work. There is no real difference if we are given the education, training and empowerment to be able to make a difference.

Pause for thought

Before moving on, just think about the work you do and ask yourself, 'Is there something that I (or we) would like to be seen as being better at and I/we would like recognition for?' It does not have to be anything big. Even the very small things are important, especially if they make us feel better and we keep on doing them. This is often referred to as being 'incremental improvement'. Every day, try to do things better than the day before! It is a good feeling. Make a note of it in your notebook and leave space for more ideas that occur to you later.

Figure 1.1   Cycling team track training

However, it is not just in doing our job that we want to understand how to achieve quality. We needed to have some idea about 'quality' to get the job that we have in the first place. The interviewer is looking for signs of 'quality' in you! When we see an advertisement, do we know the best way to respond? If we were unsuccessful, do we know how to find out why and learn what to do to get a better result next time? You see, 'quality' is involved in everything that we do.

The great thing about understanding how 'quality' can be achieved is that it can and should be life changing. Being the best individually and collectively is just great! Ask the members of Team GB after they came home with their bunches of medals. This could be you and your colleagues and friends in whatever you choose to do. That is what this programme is intended to help you do.

Bear in mind – you cannot 'standardise on being the best'!

Regardless of who you are and whatever your role in an organisation, *you* are personally responsible for your own quality if you are given the chance. At least, you are responsible for the quality of what you do. Nobody else, just you! So, a question you might well ask is if this is the case then why do we have 'quality managers'?

Why indeed! There is a good reason though: in the late 1950s a well-known American quality guru, Dr Feigenbaum, observed that 'Quality is everybody's business'. Unfortunately, whilst he was partly on the right track, he did not think this through and he went on to say that because it was everybody's business, it could become nobody's business, so he went on to describe the function of the so-called 'quality department'.

This, in turn, resulted in the systems that we see in the West today where 'quality management' has become a police-style activity – Whose fault is it? – the result is often 'a blame culture' in which people become alienated from 'quality' rather than supportive to it. When something goes wrong the question then is 'Whose fault was it?' This is crazy because nobody goes to work with the idea of doing a non-quality job. This is very different in parts of the Far East.

In Japan, the reaction to Feigenbaum's original observation was very different. Instead of developing a 'quality department' function, they set about organising 'quality' in such a way as to involve everyone at all levels in the accomplishment of corporate quality goals, with everyone being responsible for quality at their own level. It might seem surprising but most Japanese companies do not have a 'quality department' as such. Instead, they have either steering committees or quality councils made up of representatives of all key functions. More on this later.

Here is an extract from a Japanese periodical: 'In Quality Control work, all persons concerned from the President down to the operators have received suitable education and training. In one word, we have practice, so to speak, quality work of, by and for all.' This is the essence of *Hoshin Kanri*.

Pause for thought

Before moving on, how much education and training in the Quality Sciences and disciplines have you and the employees in your organisation had or are having? Highlight this as an open question and return to it as you progress through the book.

So, if an organisation were to adopt this approach would it mean the end of the road for the quality manager?

From his/her point of view, fortunately not, provided that he/she has a broad and in-depth knowledge and understanding of the Quality Sciences and disciplines. However, it should result in a significant role definition change in many cases.

It will take a long time for most organisations to realign everyone's thinking from seeing the quality professional as being a policeman and, instead, to being facilitator, coach and mentor but more on that in a later chapter. Their job then will be to assist those who really are responsible for quality, to understand the concepts relevant to them and then to be able to practise them in the workplace.

To be respected by all levels of management from the chief executive through to the most basic level by his/her personality should be a key goal.

Also,

- Caring/sharing
- Good listener
- Good teacher
- Team builder and skilled in team dynamics
- Enthusiast for what he/she is doing
- Enthusiast for the achievement of the goals of the organisation
- Enthusiast for people building and their personal development
- Team player
- Belbin characteristics – some aspects of shaper but not excessive, monitor evaluator, some completer/finisher
- Nothing is too much trouble.

Earlier, we stated that 'quality' is contextual, implying that its definition depended on the situation. Over the years many experts have attempted to define it and here are some of the suggestions.

Nearly all of the so-called quality gurus have had their own favourites.

- Dr Juran used to speak of fitness for use
- For Phil Crosby, it was 'customer satisfaction' then there have been others such as 'according to specification', etc.

Even today, the same arguments prevail and the differences of opinion as to a precise definition of 'quality' remain as rigid as ever. Personally, we are happy with our own definition: **'quality is the sparkle that separates the best from the rest'**! The ISO 9000 standard defines quality as the 'degree to which a set of inherent characteristics fulfils requirement'.

- **Manufacturing**: strict and consistent adherence to measurable and verifiable standards to achieve uniformity of output that satisfies specific customer or user requirements
- **Objective**: measurable and verifiable aspect of a thing or phenomenon, expressed in numbers or quantities, such as lightness or heaviness, thickness or thinness, softness or hardness
- **Subjective**: attribute, characteristic, or property of a thing or phenomenon that can be observed and interpreted and may be approximated (quantified) but cannot be measured, such as beauty, feel, flavour and taste.

Sometime towards the end of the 1980s, Dr Juran came up with the idea of what has become known as the big 'Q' and the small 'q'. The **small 'q'** refers to each individual product or service whilst the **big 'Q'** refers to the reputation of the organisation as a whole in its market place.

For example, the reputation of, say, Toyota as a corporation rather than simply the performance of its production processes. They are, to some extent, mutually dependant but there are many issues that impact the big Q that are not related to the small q, for example investment policies, employment policy, marketing and sales, etc.

Figure 1.2   Dr Juran

Pause for thought

If you had to define the meaning of the word 'quality' now, what might you say?

In this section you will have learned:

- Who is responsible for 'quality' (who do you think it is now?)
- What 'quality' is
- Why a 'quality manager' (or why not!)
- Role of the quality manager
- Expectations of the quality manager
- Specific definitions of 'quality'
- Some additional definitions of 'quality'.

SECTION 2: A BRIEF HISTORY OF QUALITY MANAGEMENT

In Section 2 you will learn:

- A brief history of quality management
- What is craftsmanship?

- What is mass production?
- Quality inspection: at what point is it carried out?
- Reasons for product failure
- What are self-managing teams?

Surprisingly, 'quality management', as such, is a relatively new phenomenon and the title was only used in a very small number of companies as recently as the 1960s. In fact, it only came into common usage in the early to mid-1970s when the Ministry of Defence published the so-called Defence Standards for Quality Assurance.

Prior to that, it was more usual in the West to rely on the 'chief inspector' to stop poor quality from reaching the customer. Unfortunately, a lot of it still did. Let us look at the history of this to find out how and why the changes came about.

As commented in the Introduction, prior to World War I, there would have been many companies that did not even have an inspection function, let alone quality management. The reason for this was that the company would have employed **'craftsmen'** – workers skilled in a craft.

Such people would have completed a very tough apprenticeship, usually under the watchful eye of a highly skilled individual. Typically, the apprenticeship would have lasted five years but in some cases even longer. During that time pay would have been very poor and, in some cases, the apprentice's father would have to pay for the young person to get the job.

In the UK, it was more the impact of World War I that began the change. Sadly, most of the skilled craftsmen were killed in the trenches and mud in France. In the factories, their places were taken by largely unskilled female labour and, consequently, it was necessary to introduce the inspection system.

Previously, there had been no need for quality control as the craftsman was more than capable of checking his own work. However, prior to the changes in the UK, all of this changed in the USA in the early part of the twentieth century. In their case, this was due to the introduction of the 'division of labour' concept or mass or flow line production as it was referred to by some.

Mass production involves making identical products very quickly on an assembly line. Workers each work on part of the product rather than one worker working on the whole product from start to finish. The unskilled workers spend little or no time preparing materials and tools so the cost of producing a product is greatly reduced.

The quality aspect was assigned to the quality assurance department. This department, using various techniques, checked the quality post-production. It resulted in the creation and use of statistical process control or statistical quality control as it was known through to the late 1950s.

In mass production, inspection is a post operation activity and not generally carried out by the workers themselves. In this case, it is usually carried out by a person who may not have the skill to do the work but hopefully does know what the result should be like.

In a craft environment the inspection takes place during the operation and is done by the craftsman himself as one would if painting a door or something like that.

It has been found that over 90% of inspection is visual and 80% uses no artificial aids of any kind. It is also true that human beings, regardless of skill, intelligence, motivation or personality, miss a minimum of 10% of defects for no other reason that that they are human! So much for 100% human inspection!

In another chapter it is shown that sampling inspection when done scientifically is lower risk than 100% parts inspection. in the case of highly repetitive activities such as the visual inspection of high-volume products. The reliability of the human inspector is a highly complex topic and is rarely referred to within the quality profession. We are not covering it in this book but for those who are interested, and want to research the subject as there is little published material on the topic, useful words to google are 'time on task', 'vigilance decrement', 'signal to noise ratio', 'multiple monitors', 'Broadbent's single channel theory', 'Mackworth Clock Test', 'the reliability of the human inspector by Drury and Fox', 'Harris and Cheney', and from these will follow physical factors, environmental factors, workplace design, etc. Much of the work is PhD research level, delves deep into human psychology and not entirely intelligible to most, but there is much to be gleaned from reading the more straightforward aspects if you pick through the jargon.

In the mid-1960s, it was realised by some that it did not matter how much you inspected a bad product, it would not make it any better and all inspection would do to a good product is to add cost (it is better to attempt to make it right in the first place). This gave rise to the 'right first time' or 'zero defects' movement.

Dr Juran, the world-leading quality guru at that time, stated that this movement was largely a failure due to erroneous premises. It assumed that the cause of failure was down to poor motivation. All that was necessary, therefore, was to 'motivate the workers'. The fact is that there are multiple causes of product failure and the motivation of the workforce was only one of those possible causes. This caused a great deal of resentment amongst the workers.

Figure 1.3   Professor Kaoru Ishikawa

Even earlier than that, in Japan, Professor Ishikawa, a very perceptive quality expert, observed that the cause of the problem was in the way that people were treated, and he postulated a revolutionary solution. Involve the workforce! In effect, he advocated 'reinventing the craftsmanship system', not on an individual basis because that was uneconomic, but involving the workforce in self-managing team activities.

As a consequence, they had no need for the armies of inspectors because for the most part the workers took control of the inspection of their own work, kept records and found the causes of problems and implemented remedies. These teams were called 'quality circles', in general, but some people, especially Nissan, call them Gemba Kaizen activities as an alternative, but the two concepts are identical.

Pause for thought

How much do you know about the efficiency of 'inspection' of your products and services? How dependent are you on this activity as a means of ensuring that your defects are detected as soon as possible? Can you think of any possible alternative at this stage? Fortunately, some will be suggested later.

Summary of learning

Historically, it was the quality inspector who would try to prevent poor quality reaching the customer. It did not matter how much you inspected a bad product; it would not make it any better, and all inspection would do is to add cost to the product.

More recently, originating in Japan, there have been big changes to how quality is managed. Workers are encouraged to take control of the inspection of their own work and to implement remedies. These teams of workers are called 'quality circles'.

SECTION 3: THE COST OF POOR QUALITY

In Section 3 you will learn:

- Defining the cost of poor quality in manufacturing and service industries
- Better quality-related costs more or less?
- Quality-related costs in service operations
- Quality-related costs – general
- Unnecessary perfectionism.

We are sure that everyone has at some time or other said 'well I cannot afford better quality', the supposition being that better quality costs more. Well, we may have a surprise for you. In many cases better quality costs **less** not more! It all depends what you mean by better quality! This is a perennial debate, it depends what you mean by 'quality'. Even within the context of the definitions in Section 1, there can be widely differing opinions on whether quality costs more or less.

If you are talking about the **'quality of specification'** then definitely 'better quality' costs more. There can be no doubt that a gold-plated pen costs more to produce than a simple throwaway ballpoint. This is what some people erroneously think is 'quality'; however, this belief does not consider **'fitness for use'** or **'value for money'**. As pens, they both perform the same function. If the pen is intended as a birthday present for a loved one, the plastic ballpoint would hardly be adequate!

We have two companies both producing gold pens to the same specification. One is 'The Perfect Company Ltd' that never does anything wrong. The other is the 'We Somehow Get By' company. The company's manufacturing plant typically has excess inventory, scrap, rework, long set-up times, multiple product design changes, etc.

The Perfect Company Ltd that never does anything wrong has perfect raw material that comes in at one end, exactly on time when required. Each element of the manufacturing process is perfectly set exactly on time. There are no hold-ups or delays, no scrap, waste or other losses and a truck is waiting to take the finished product to the market.

In this organisation, provided that the pen does what is intended and meets all the requirements of the customer, there will be no quality-related costs. Now we have the contrasting situation, our Somehow We Get By company. There are late and sometimes wrong deliveries of raw materials, processes are stopped, and materials are returned due to problems, so we have stock-outs.

There are stocks of goods to be processed, tying up much needed capital. There are unscheduled stoppages and so-called 'work in progress', machines running slowly due to poor maintenance, poor raw materials and, eventually, there are huge finished goods stocks due to poor scheduling.

In Figure 1.4, which of the following two companies has the lowest cost to produce?

So, which of the two companies is the most competitive and which is likely to produce the most consistent and high quality (in terms of meeting the specification requirement) of the two? No prizes for guessing!

As we said above, sometimes we are talking about the quality of the **specification** and not always about the 'quality' of the product. Here is another illustrative example: You are in a supermarket choosing a bottle of wine. On the shelf, at one extreme there is **'Cheapo Plonko'** at, say, È4 per bottle. And at the other extreme there is **'Classico Fina Wina'** at È22 per bottle. So, which of these do you expect to be regarded as the best quality? Cheapo Plonko or Classico Fina Wina?

Well, we guessed you would say Classico and, of course, you would be right and, yes, in this case better quality costs more. This is because the **specification** is more exacting.

The chances are, though, that on an everyday basis most people would buy Cheapo Plonko but, on special occasions, might opt for Classico, for example, a wedding celebration if we could afford it!

Competitive quality-related costs

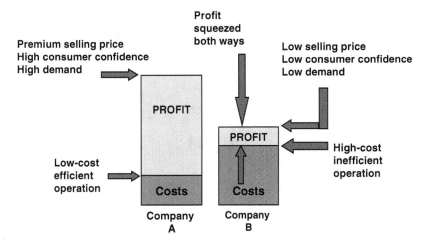

Figure 1.4   Quality-related costs

The lesson is that if we are talking about quality just in terms of **value for money** then probably, we would say Cheapo Plonko is the winner. Both wines are probably pleasantly drinkable, but one is four times the price of the other because it is made to a higher **specification**.

However, whilst they are both wines and sold from the same shelf in the supermarket, for a special occasion such as, say, a wedding or anniversary, we bought Classico Fina Wina **because** it was more expensive, and we wanted our guests to think we have lavished some money on them. Paradoxically, in this situation, if we reduced the price of Classico Fina by 50% it might not result in extra sales because people were buying it **because** it was expensive! In this sense, Classico was the better value for money and more fit for this particular purpose.

Let us now imagine that we have two wineries and **both** of them are licensed to produce both Cheapo Plonko and Classico Fina Wina. One of them is the Perfecto Company, which is very well organised, and the other is the Letsope for the Best Wine Company, which somehow just about survives. Both of them produce the same wines.

The Perfecto Company

In the Perfecto Company, as a consequence of their perfect planning processes, perfect grapes come in precisely at the time expected. They go straight to the first operation in the production process, which has just been made ready for them and which is just perfect. They go through the fermentation process exactly on time, with no hold-ups, no delays and no 'in-process' wastage. Eventually, they are bottled and sent immediately to the market on a truck that is ready and waiting so there is no 'finished goods' warehouse – they do not need one. Cost of

Figure 1.5   Stock insect damage

wastage, nil, cost of delays nil, in-process losses none. So, cost of poor quality, none and **everything** was added value!

The Letsope for the Best Wine Company

In the Letsope Company, things are different. The grapes arrive unexpectedly, sometimes before the process is ready, so they have to be stored and, as a consequence, some of them start to rot and others are damaged by wasps and other creatures. On other occasions, the grapes arrive late with the processes standing idle. How much does all of this cost?

The plant is not in good condition so is subject to many breakdowns, slow running, and out of specification product.

Pause for thought

- How much do all of these non-added value problems cost?
- Delivery promises are made and broken and, in the end, much of the wine must be stored in a finished goods warehouse due to unpredictable delivery schedules.
- How much of the company's capital is tied up in this form of waste?

- Guess which of the two companies has the highest quality-related costs to run?
- Where do you think your organisation might be in relation to its competitors if the truth were known?

Summary of learning

We can see from the above example that if we talk about 'quality' in terms of the 'specification', then clearly better quality does cost more money. However, if we are talking about 'quality' in terms of the way we do things around here, then better quality costs less! The better the quality of our processes and procedures, the lower the cost. So, the argument 'we cannot *afford* to produce better quality' *only* applies to the specification and *not* the process.

SERVICE INDUSTRIES OR SITUATIONS

We have just looked at a manufacturing example. Let's now look at a service industry situation. Quality-related costs here are not so obvious but that does not mean they do not occur. In fact, studies show that in many cases they are far higher than in manufacturing for just that reason. For example, the daily activities of a secretary, travel courier, nurse, surgeon or salesperson may not be recorded, and the breakdown of time spent on each activity may not be known even to the individual concerned, other than by unreliable, subjective means.

Even when timesheets are used to record time spent on various activities, it does not follow that the time was used efficiently. A secretary is unlikely to record that three hours were spent looking for a file, if the location of it was the secretary's responsibility. Also, in service industries, as distinct from manufacturing, **the customer is part of the process and affects the quality of the process**. A polite customer will often receive a different quality of service from an impolite customer. It is a golden rule not to upset a waitress because you have no idea what she might do to your food in the kitchen!

Pause for thought

Quality-related costs in your workplace

For your organisation's products or services, think of an example where the quality of the product might either be 'over' or 'under' specified.

Unnecessary operations in manufacturing in many cases will require additional plant to carry out these unnecessary operations and this must be justified financially. A machine operator's tasks are usually precisely defined and leave little scope for variation. Whichever term is preferred for waste, it includes all costs and activities that do not add value. The cost of lost business because of the customer experiencing our problems is also worth considering but is unfortunately not quantifiable. However, it is usual that when successful attempts are made to reduce cost of poor quality, there is generally a significant increase in business. **Unhappy customers vote with their feet!**

This fact applies as much to the service industry as it does to manufacturing organisations; in fact as we have said, experience indicates that these costs are often much higher in a service industry because they are less visible. For example, the consequences of many manufacturing problems result in excessive inventory, process downtime, customer returns and warranty costs, late payments due to disputes over quality quantity and delivery. These are usually highly visible.

Having argued the case for quality-related cost reduction, the student will want to be able to identify quality-related costs in his or her own organisation. However, it has already been mentioned that many of these are hidden in the accountancy process. Fortunately, there are ways that can be used to pull them out. One way is to use the hidden factory concept first described by Dr Feigenbaum in his book *Total Quality Control*, first published in the 1950s.

The idea behind the hidden factory concept was that every factory, office, shop or operation of any kind consists of two factories, offices, shops or operations of any kind. One of them does nothing wrong just as in the case of the Classico Wine Company earlier. The other one does nothing right. It is even worse than Cheapo Plonko. Of course, in reality, these two companies are integrated together under one roof.

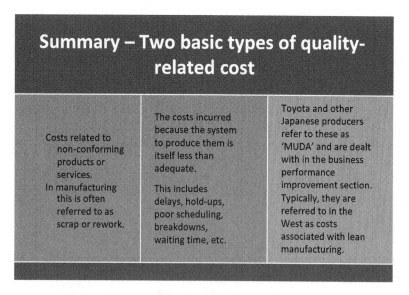

Summary – Two basic types of quality-related cost

Costs related to non-conforming products or services. In manufacturing this is often referred to as scrap or rework.	The costs incurred because the system to produce them is itself less than adequate. This includes delays, hold-ups, poor scheduling, breakdowns, waiting time, etc.	Toyota and other Japanese producers refer to these as 'MUDA' and are dealt with in the business performance improvement section. Typically, they are referred to in the West as costs associated with lean manufacturing.

Figure 1.6   Types of quality-related costs

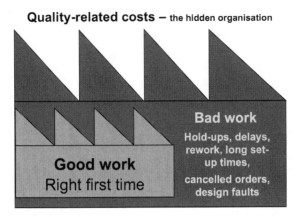

Figure 1.7   The hidden factory (or office)

The question is, how much does it cost to run the company that does everything wrong? It is important to have at least some idea because this is the real cost of your poor quality!

In this concept the above diagram shows that an organisation really consists of two organisations. In one of them only good work is produced. In the other, there is only rubbish, waste, delays and errors. In this latter organisation how many people does it employ? How much floor space, how many machines or pieces of equipment are used just to produce the bad product?

In Figure 1.4 we can see the difference in the market situation of Company A, which is close to being perfect, and Company B, which is at the Cheapo Plonko end of the spectrum. If Company A decided to bring its sales price down to below the cost base of Company B then Company B will not be price competitive and, as a consequence, it could mean an exit from the market for Company B.

Figure 1.8 gives a few examples of the quality-related costs in the hidden factory. Note that they exist in different forms in all departments. Most of them never appear in the accounts but are absorbed into blanket overheads. If we do not account for them, we will never know what impact they may have, we will continue to ignore them, and they will continue until the company fails.

The argument thus far assumes that improvements in quality are perceived by the customer to be desirable, and that the customer is prepared to pay for them. Quality improvements that cannot be perceived by the customer, or to which the customer is indifferent and cost more to achieve, or for which the customer is not prepared to pay, must be regarded as unnecessary perfectionism and should be avoided. These improvements should be regarded as quality deficiencies just as much as those that relate to failure to achieve. Sometimes unnecessary perfectionism can be one of the most significant quality-related costs.

One example of this was the manufacturer who had a near absolute monopoly of the smoke detector market. At that time, soon after clean air laws had been introduced, the company was able to sell its products at premium prices because it

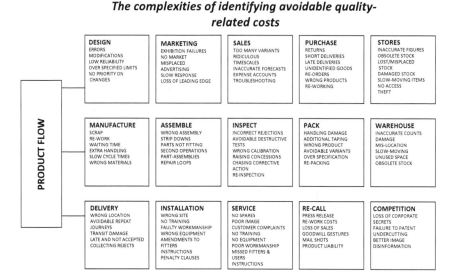

Figure 1.8   Avoidable quality-related costs

had no competition. However, the company had also fallen into the trap of unnecessary perfectionism. The directors insisted that the outer cases of the product should be die-cast and vitreous-enamelled to look nice and to be consistent with the appearance of its other products. These design features were extremely costly to produce and were totally unnecessary from a practical point of view.

A perceptive competitor entered the market with a product with far more utility. It was equally efficient but half the price of the other and therefore captured the market in no time.

Pause for thought

Although it is often stated that the cost of poor quality is often at least 30% sales revenue, this is often shrugged aside but do not do that. Think about it. Supposing it is true for you! Unbelievable? Well, many have thought that, only to be dramatically proven wrong. Do not rush out and hire an expensive consultant to check it. That will only add to the cost. You can do it yourself easily. As you work through this book, you will begin to see why, if you cannot already.

But you might then say, well suppose it is that high. That does not mean that we can do anything about it! Well, you could be the exception but, generally, organisations that challenge this generally find that they can halve the cost inside three years, in many cases much better than that. Just imagine. Suppose your turnover is È10m per year, with an annual profit of 10% of that, and to keep the numbers easy we assume 20% quality-related loss. If we can halve it, that represents a doubling of annual profit. What other means do you know that could achieve that without some huge investment and risk taking?

Summary of learning

Quality-related costs occur in the service as well as the manufacturing industry but are not so easy to spot. Quality-related costs are all costs and activities that do not add value and the cost of lost business as a consequence of the customer experiencing problems.

SECTION 4: QUALITY PLANNING

In Section 4 you will learn:

- Quality planning is an essential activity that commences as soon as a new concept has been defined and is intended to be developed.

It begins (or should begin!) with the evaluation of whatever market research information may be available and runs through all of the proposed activities of conceptual design, functional design, design for operations, the 'make or buy' decision and the relevant decision criteria, vendor evaluation and control, internal operations, packaging, shipping and delivery, after sales product support and, eventually, through to final disposal at termination of its useful life. This can be a very long process in some cases and may even go beyond the expected working lifetime of many of those involved.

Quality planning is an essential activity that commences as soon as a new concept has been defined and is intended to be developed. Please study Figure 1.9 to see the basic elements.

Market research

Market research is the activity carried out to determine market opportunities and potential customer needs. It is highly specialised and usually involves considerable statistical analysis of demographics.

Conceptual design

Conceptual design supposedly carries out analysis of market research data in order to create designs to meet the perceived need.

Potential quality problem!

Designers are highly technically capable and sometimes are more interested in promoting their own ideas than those resulting from market research.

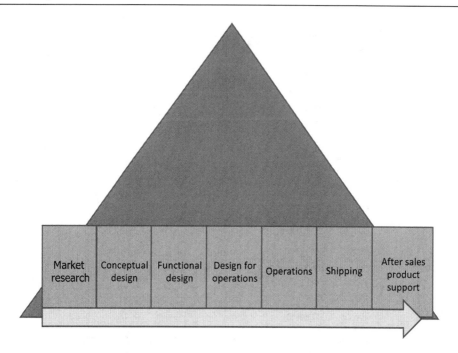

Figure 1.9   Key business processes

Functional design

With complex manufactured products or service situations, Functional design might require considerable design proving in order to ensure the product or service does not contain glitches that are not immediately obvious.

Potential quality problem!

There is the possibility of hugely expensive product recall and risks to life and or the environment.

Design for operations

This includes the 'make or buy' decision. We need to carry out pre-production runs to be sure that we can produce the product or perform the service economically and that the trained people and facilities are in place before production.

Potential quality problem!

Design throw the product over the wall into production who find it is impossible to produce. Massive costs and delays may result from design change requests.

Operations

This includes supply chain management. In a perfect world we get perfect products if people are properly trained and developed, given the opportunity to manage their own work, have the means to do good work and know what is expected of them.

Potential quality problem!

People may not be properly trained, may not have 'self-control', or are not properly supported and facilities are often inadequate, so problems abound.

Shipping

In an ideal world with perfect design and perfect operations we should also have perfect scheduling, therefore there is no need to keep expensive finished goods stocks and products can go directly from production to the customer.

Potential quality problem!

We do not live in a perfect world. Scheduling problems, long set-up times and plant breakdowns, etc. result in stockouts, excess inventory, obsolete or deteriorated stocks and high space rental costs.

After sales product support

If we have designed and made a perfect product or service, we want the customer to get maximum satisfaction from it. This includes customer instructions.

Potential quality problem!

We might skimp on the design of instruction manuals and type them in an unreadable small point size, the packaging is often difficult to remove. The customer fails to get satisfaction because he/she never gets to know how to use all its features.

From the foregoing it can readily be appreciated that approximately 80% of the cost of poor quality has its roots in the design stage. If poor quality has been designed in then no amount of manufacturing care or inspection will take it out later, therefore that aspect of quality planning is critical to success. Between each of the key stages in the design process, it is necessary to thoroughly review all quality critical features before progressing to the next stage. This is an iterative process because if some things are not quite right, they must be repeatedly readdressed until they are.

Figure 1.10 shows a typical quality planning process based on the forgoing argument. The GAPS show where the biggest and most likely problems arise.

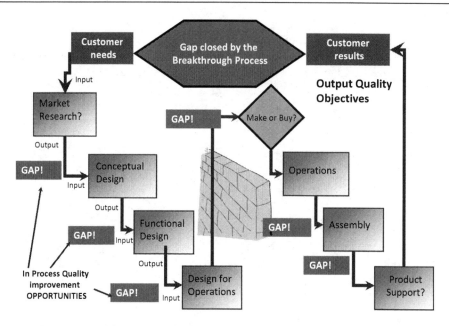

Figure 1.10   Problems with silo management

Pause for thought

Before moving on, try this. Get a group of say four to five people from any function that you like then select a similar number from an upstream or downstream department.

Ask one of the teams to carry out a brainstorming session in a separate room from the others, which lists all the things they receive from the other department. Then make a separate list of all the things they send to that department because traffic often goes in both directions. Then ask them to order the items on each list in their perceived order of importance. Ask the other team to do the same. After this, ask them to compare their two lists. I would love to be a fly on the wall when this happens! You will probably be amazed. There will be some issues you can be sure!

What problems can you see that result in everyone blaming everyone else?

There will also be issues such as:

- People in the operations department saying, 'If the designers knew anything about manufacturing they would not have designed it that way'.
- Response from the design department: 'If you were good at manufacture you should be able to produce everything we design'.

Summary of learning

Quality planning is an essential activity that commences as soon as a new concept has been defined and is intended to be developed.

It begins with the evaluation of whatever market research information may be available and runs through all the proposed activities of conceptual design, functional design, design for operations, the 'make or buy' decision and the relevant decision criteria, vendor evaluation and control, internal operations, packaging, shipping and delivery, after sales product support and eventually through to final disposal at termination of its useful life.

SECTION 5: PRODUCT RECALL

In Section 5 you will learn:

- Product recalls can happen in all areas of industry
- Who is responsible for overseeing a product recall?
- When do you initiate a product recall?
- Product recall procedures
- Types of product recall
- Hazard evaluation report
- Warning notice.

Workplace thought! Have you ever experienced a product recall?

If you are a car owner, the probability is, yes, almost certainly. They happen very frequently in most models and most makes. But it is not just cars. It can be anything from baby foods, steam irons, washing machines, mobile phones, in fact, anything. We even know somebody who experienced one with a heart pacemaker! Not exactly easy to send it back! Why should this concern us here? Sometimes, the organisation making the recall knows who and where you are as is the case with automobiles. In this case, you might receive a letter asking you to return your vehicle for a free replacement of some part or other. In other cases, the manufacturer cannot locate you as would be the case with baby foods, in which case it would be forced to use TV and other high-profile advertising.

As we asked, why should we be worried about a product recall here in this introductory programme? The reason is that the cause of a possible recall could be at any stage in our processes. Sometimes, the problem is minor and only affects a single batch of product. At the other extreme, the problem might be life threatening to a lot of people requiring the highest level of alarm to avert a disaster.

The challenge is not to allow a product recall to threaten the entire brand or company. Research indicates that negative news is devastating; on average, the media impact of negative news has quadruple weight when compared with positive news.

- Intel's 1994 Pentium microprocessor recall allegedly cost $500m alone.
- Coca-Cola posted a 21% drop in income linked to its European contamination scare and recall of 17m cases of Coca-Cola from five countries in 1999.
- Firestone's Ford-related tyre recall is estimated since 2001 to have wiped out more than half its parent Bridgestone's profits and dragged down the share prices of both companies.

The impact of negative news can be made worse by a slow response to the problem.

Does it only apply to manufactured products? What about services? Well, quality-related problems are just as common in services as in manufacturing. It is more likely that we must make sure that whatever it is that people are doing that is risky is either stopped or modified in some way or some warning notices might be given.

Maybe we sell holidays. It could have overlooked some important piece of advice or maybe something that was advertised has changed and we need to inform our customers immediately.

The reason why we need to be worried about a product recall here in this introductory programme is because it does not matter how careful we try to be, accidents do happen and with the best will in the world things do occasionally go wrong. In fact, thousands of things go wrong but most of them do very little harm other than put up costs a little and slow down the processes but, occasionally, it is worse than that and some unwanted failures might be life threatening. So, just as we have fire extinguishers and first aid kits, we should also be ready for the possibility of having to recall our products from the market if something has gone wrong.

Like heart attacks, cancer, road accidents and muggings, product recalls happen to other organisations but could never happen to us! That is what everyone thinks until it does happen and when it does, we may have to act fast.

Figure 1.11   **Quality-related accidents**

PRODUCT RECALL

PATAK'S JALFREZI 450G COOKING SAUCE, WITH
BEST BEFORE END (BBE) CODE 12.2014 AND
NUMBER 3178 PRINTED TOGETHER ON THE JAR LID

Patak's is recalling the above product as a precautionary
measure due to a quality issue.
No other Patak's products are affected

WHAT YOU SHOULD DO

If you have purchased this product, please contact the
freephone number **0800 197 1679**.
Freephone number for Eire **1800 928 188**.

We apologise for any inconvenience caused and would
like to thank all our customers in advance for their
understanding and co-operation.

FREEPHONE NUMBER
0800 197 1679 (UK)
1800 928 188 (Eire)

ONLY APPLIES TO
PATAK'S JALFREZI 450G COOKING SAUCE,
BBE CODE 12.2014 & NUMBER 3178
PRINTED TOGETHER ON THE JAR LID

Figure 1.12   Product recall notice

Product liability situations can arise at any time and almost always without warning. It can, and does, happen to the most quality-conscious organisation as well as the careless. No organisation could ever say with absolute certainty that it could not happen to them if they manufacture or distribute products in the market place.

Can we insure against it? Yes, sometimes but the higher the risk, the higher the premiums if, indeed, insurance can be obtained at all. The risk can be reduced but not eliminated through the application of quality-related disciplines. A product recall procedure can greatly reduce the costs and minimise the effects if the worst happens.

A product recall will always be the result of the discovery of a real or potential hazard, which will require the fastest possible location of every item responsible for the risk and its removal from the market. Experience indicates that, with the best will in the world, this is close to impossible.

A product recall plan, to be effective, must be capable of instant implementation. Decisions may need to be made on the recall of products from overseas, the tracing of individual purchasers, television and newspaper advertising, social media communications, the withdrawal or freezing of stock in warehouse and retail outlets, and stock replenishment.

The ability to rapidly issue instructions for the safe use of the product or its disposal, the subsequent monitoring of the health of those who may have inhaled or consumed doubtful chemicals, foodstuffs, gasses, radiation, etc., not to mention the

Product recall item

Product recalls are made by traders about products that have problems which could affect the safety of the user/consumer. The product should not be used and should be returned to the trader.

Traders wishing to display their recall / safety notice on this page should email it to publications@tsi.org.uk. Latest RAPEX notifications are available on the Europa website. Vehicle recall notices can be found on the Driver and Vehicle Standards Agency website.

John Lewis Wooden Shape and Colour Sorting Board - recall

The following notice has been issued by the retailer:

John Lewis Wooden Shape and Colour Sorting Board

The safety of our customers is very important to us, so as a precaution we're recalling the John Lewis Wooden Shape and Colour Sorting Board from sale. This is because we have identified a potential issue during the manufacturing of these items that may cause the ball on the blocks to become detached, which presents a choking hazard. We have had no reported incidents, but are recalling this product as a precaution.

Figure 1.13   Typical product recall announcement on the internet

internal costs of product modification, replacement and the administration of the plan, is critical if the worst consequences of a product recall are to be avoided.

Should a recall be required, speed of action is of the essence. This is not the time to have long discussions as to who should be in charge and to direct the plan. This should have been decided long before such an event. Often, but not always, this responsibility should fall upon the shoulders of the quality manager because he will be the most familiar with the systems in operation. However, depending upon the situation, all of us could be involved! It couldn't happen to you? Well, think about all the possible things that can go wrong because someday just one of them might!

The quality manager may also be involved in leading the resources assigned to deal with the media should the need arise. In the case of urgent or serious product recalls, which may be costly, likely to damage the company's reputation, or receive any form of national or local publicity, it should usually be necessary to obtain board approval for any announcements or interviews that may take place.

In summary, a recall should always be initiated when it has been ascertained that there is a definite or potential danger to life or health, and when the continued use or circulation of the product is likely to result in legal action. A recall may also be initiated when it is found that a fault in the product, or lack of performance is likely to affect the reputation of the company and when the lack of performance fails to justify the claims made in advertisements. This would also include drugs, where the dosage advised was found to be at variance with requirements.

Pause for thought

Can you think of any possible situation where your company's product or service might need to be either recalled or stopped with high publicity to prevent risk to life?

Do you already have measures in place to deal with it? How much might it cost in the worst-case scenario? Are you prepared for it?

The first stage in a recall programme will be some form of notification of a hazard. This could come from a number of sources and in many forms, depending upon the product. The most likely sources include:

- The user
- An independent test house report
- Government sources
- Own reliability test programme or research
- Research institutes and laboratories
- Overseas sources.

User level recall

A user-level recall is likely to be extremely expensive, particularly in the case of consumer products. It would be essential in all urgent cases and probably in most serious cases. It entails recall from users, retail outlets and warehouses.

Subsequently, it would be necessary to reimburse the owners for losses sustained as a result of the withdrawal. The order for such a recall would always be made at board level.

Retail-level recall

Less serious cases to do with probable law breaking, particularly in cases where an existing product has been in widespread use for some considerable time, and now information has revealed a minor hazard.

Wholesale level recall

This will entail product recovery from the first stage in the distribution chain and may be applied in the case of non-safety critical situations

Limited recall

Limited recall means the recall of individual batches and consignments and will probably occur most often when manufacturing faults have been discovered or are suspected in an established product.

Hazard evaluation report

This evaluation should be carried out by the manager responsible for the recall programme. In any case, it is recommended that a hazard report should be prepared and circulated to all concerned within two days of instigation, regardless of the action taken. The report should include a description of the type of hazard, the category, and the recall level.

Apart from good housekeeping, hazard reporting provides a good source of data for future risk probability evaluation. As with accidents, near misses are far more numerous than actual events. By recording near-miss recalls, the company will soon be able to evaluate the likelihood of an actual event and its possible consequences.

Upon receipt of the agreement by the chief executive or his nominee to initiate the recall programme, a hazard warning notice may be required. This should contain all the information necessary to bring the risk to the attention of all those likely to be exposed to the hazardous product and the circulation may include:

- Field specialists including service staff
- Appropriate government or local authority departments
- Police and other emergency services
- Press, local and national where applicable
- Radio and/or TV, local and national where applicable
- Retail outlets
- Customers, where possible
- Trade journals and trade associations
- Professional institutions and research establishments
- Export agencies/licensees, foreign governments, etc.
- Relevant test houses.

A warning notice is a statement of the nature of the hazard for example, electric shock, fire, explosion, inhalation of noxious gases, etc. and the likely consequences.

- Clear instructions and illustrations to enable the repair, correction, removal, immobilisation or shielding of the product
- Warning or hazard signs, which may be detached from the notice and attached to the product, and its container
- Instructions appertaining to the return of the product, and a recall return card, preferably reply paid. This could take the form of a tear-off portion of the letter.

Communication with the customer

The internet has added a new layer of visibility to product failures and product recalls. Social media can make or break a company in crisis, but it can also demonstrate a company's regard for its customers.

Open communication channels

Communication channels include a customer service line and a website with details and instructions for customers. A mishandled public response can cause more damage than the problem it addresses in the first place.

Once the recall procedure is under way, it will be necessary to assess the effectiveness of the procedures adopted and to consider the possibility of stepping up the campaign. Similarly, it will also be necessary to make a decision to terminate the recall when all possible items have been returned.

Due to changing international legislation, the number of major product recalls has been on the increase, and now is a major source of worry to industry. Quite apart from the actual costs of managing such a project, the damage to the reputation of the company can be lasting. Also, there is the question of negative advertising. Most organisations spend enormous sums attempting to convince the world that they are the best. Now they are having to spend the same sums of money explaining that their product is quite harmful. Even with the most intense application of quality related concepts the risk is never zero!

Summary of learning

A product recall should be initiated when there is a definite or potential danger to life or health, when the continued use or circulation of the product is likely to lead to legal action and when the fault in the product is likely to affect the performance of the company. The impact of negative news can be made worse by a slow response to the problem. There are different levels of product recall: user level, retail level, wholesale level and limited recall. A hazard report should be circulated to all those involved in the product recall. It provides a good source of data for future risk probability evaluation.

SECTION 6: QUALITY CONTROL

In Section 6 you will learn:

- The origin of quality control
- What is quality control?
- How quality control differs from quality assurance.
- The role of the quality control function.
- What are the typical activities of quality control?
- The types of quality problems quality control deals with.
- Quality improvement resulting from quality control.

The concept of quality control owes its origin to the work of Dr Walter Shewhart who was the executive responsible for quality management at the Western Electric Company. His ideas were first implemented at the Hawthorne plant of that company. The concept soon spread to other major American Companies.

There is a difference in interpretation of the meaning of the word 'control' in Japan and the West. The tendency in the West is to think of the word as meaning to constrain something, so to the Westerner it has connotations of using force to prevent freedom. In Japanese, due to an unfortunate misinterpretation of the term following World War II, the perception is that it means 'improvement'. This invokes some very different reactions.

However, when properly organised, both can result in quality improvement depending upon how it is organised. The reason for the difference in interpretation is that we would only want to 'control' something if it was behaving at or close to its optimal required specification requirement. Then we would wish to control it to keep it at that level. However, this is rarely the case. In most situations operations often are not running at the optimum and, generally, we are looking for opportunities for improvement and this can result from good quality control.

Figure 1.14   Dr Walter Shewhart

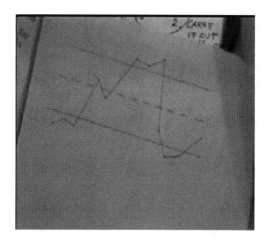

Figure 1.15   A Shewhart control chart

Typically, and in theory, quality control as an activity begins as soon as a design or specification is validated and operations to deliver either the service or product commence. The objective is:

- To have a system in place that monitors all stages of the process and to provide information to those responsible for each stage

- To be able to identify differences between actual performance and target performance
- To identify any deviation from the current state (both positive and negative) as soon as possible to make corrections and when convenient, to carry out root cause analysis on the causes of unwelcome variation.

Quality assurance is a broad companywide concept intended to achieve what the title suggests, to give assurance that the organisation will provide products or services to the level of quality expected. Quality control is a process that is set within quality assurance and provides the mechanisms that enable the overall goals to be achieved.

Once the design has been completed and the decision has been made to commence operations, the initial planning phase is over, and responsibility moves to the operating forces. The role of quality control here is to attempt to ensure that everything intended happens in accordance with the quality plan. All being well, that will be the case most of the time. However, performance will only at best be as good as the plan.

The main task of quality control is to attempt to maintain what every level of quality has been agreed and to identify and help remove any possible causes of unsatisfactory performance

Generally, it will soon be apparent that there are **two types** of problem that can emerge during this time. There will be some problems that are inherent in the process, such as waste material, unspecified stoppages, set-up losses, wear and breakdowns, etc. There will also be a myriad of people-related problems, such as sickness, absenteeism, problems due to lack of skill and carelessness, etc.

Collectively, these will result in a general level of performance lower than the planned nominal for the operation. The causes are many in number, they do not appear to be specific and usually we just learn to live with them. Eventually, we do not even notice them; they go on year after year. Also, we allow for them in the accounting methods, so even with them, there are no cost variances. Quality control does not usually concern itself with this type of problem.

In summary, quality control is more concerned on a day-to-day basis with what appear to be the more dramatic problems. The setting of a machine might drift, another might prove not have been capable in the first place. Controlling, predicting and preventing or alleviating these types of occurrence are the main task of quality control. In general, it can, or should, involve the use of statistical methods.

Pause for thought

Quality improvement in your workplace

Think of specific problems that might require quality control in your workplace.

One of the most famous of the relevant techniques developed by Dr Shewhart was the Xbar/R control chart for use with continuous variable data which could be found in mechanical processes. This data included diameters, length, weight and flow rates, etc.

The technique was found to be extremely useful for predicting trends in processes, process capabilities against specification requirements and problem diagnosis.

Summary of learning

The role of the quality function is to attempt to ensure that everything intended happens in accordance with the quality plan. There are two types of problems: problems inherent in the process and people-related problems.

The main task of quality control is:

- To attempt to achieve and maintain what every level of quality has been agreed in the specification and to identify and help remove any possible causes of unsatisfactory performance
- To use a formal approach to the analysis of performance and systematic efforts to improve it
- To use a disciplined, data-driven approach and methodology for eliminating defects.

SECTION 7: QUALITY IMPROVEMENT

In Section 7 you will learn:

- What is quality improvement?
- A typical road map for quality improvement
- The stepwise approach.

Invariably, no matter how meticulous we may have been at the planning stage, the probability will be that nothing is perfect. In fact, in many cases we will be very far from perfect. As soon as operations commence, problems abound. Some will be very serious and obvious. Others may be just as serious or even more so but not at all obvious until a lot later.

In the 1950s the UK led the world, including the United States, in the design and development of jet-powered commercial flight. The pioneering aircraft that claimed to be at least five years ahead of any rival was the DH Comet. It was a beautiful craft, and the pride of the UK.

Sadly, unbeknown to anyone at the time, it had a structural flaw that until a catastrophic failure occurred was completely undetectable. It was not until after many hours of flying time that one of the first prototypes exploded in flight killing all who were aboard.

Even then, the cause was unknown, and a bomb was suspected. Then another Comet exploded, followed by another. It was only then that the cause was discovered. The high levels of pressure change as the aircraft climbed and descended resulted in the fracture of the cabin windows causing a catastrophic loss of pressure. The result was the need for a massive redesign of the relevant components.

Figure 1.16   De Havilland Comet crash

It is clear that no matter how meticulous the planning, there is always the possibility that things will go wrong. Therefore, in addition to quality planning and quality control, we also need quality improvement.

This has two aspects:

*   The improvement of something that already exists but that requires improvement to performance
*   The development of new products and services that come as close as possible to meeting the ultimate needs of the interested parties.

In Figure 1.17, the vertical axis is 'performance in terms of deficiencies against requirement'. And the horizontal axis is 'time'. It can be seen in Figure 1.17 that, initially, the process is running at almost 20% deficiency.

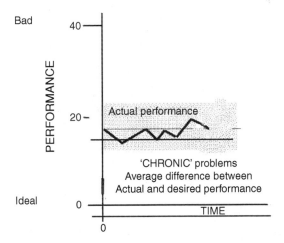

Figure 1.17   Chronic problems

Now, it may have been running at 20% deficiency for a very long time despite everyone's best efforts to make it better. Eventually, everyone accepts the 20% as a kind of norm. The accounts department allow for it in their costings. Materials purchasing allow for it when ordering; it is also allowed for in wages, scheduling and all other related activities. After that, so long as the process continues to run at 20% waste, there are no cost variances, and everyone is happy.

Suddenly, however, it jumps to, say, 40%. At that point the operating forces are alerted, and it is said that 'we have a problem'.

The question then on everyone's mind is 'what changed'? Something must have happened? This is regarded as a 'problem'. The cause is located, and life returns to normal, but the process goes right on making 20% deficiencies!

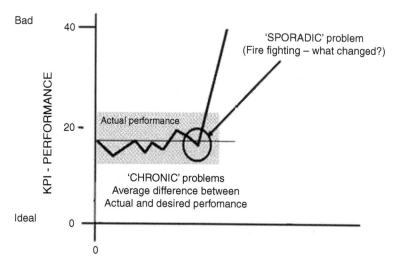

Figure 1.18   Sporadic problems

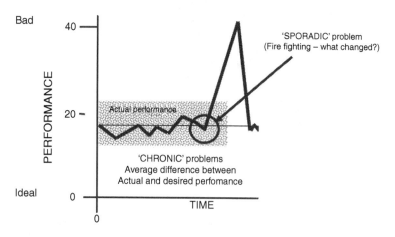

Figure 1.19   Locating the cause

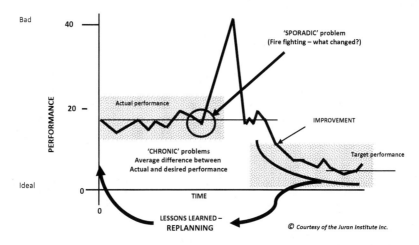

Figure 1.20   Quality improvement

So, having located the cause of the sporadic problem and applied a remedy we can see from the curve that the performance returns to its previous level. Now this is, of course, an improvement from the earlier position but we still have the underlying 20% level of waste. All we have dealt with so far are the sporadic spikes. The chronic problems remain because they are inherent in the process. To resolve these requires a different approach. We are not looking for a change. By definition, nothing has changed and we have learned to live with the problem.

What we are now talking about with quality improvement is the possible reduction of the 20% **chronic** problems to say 2%. That is a huge improvement and if achieved would go on year after year. What we learn by tackling and solving the chronic problems is a better understanding of the relationship between target and actual results. We can feed this information back to the design department, so this information is incorporated in future process design.

Pause for thought

- What might be regarded as some 'chronic' problems in your work or in your department?
- Can you estimate approximately how much they cost you?
- Fortunately, in Chapter 4 we look at the means by which these can be resolved effectively.

Figure 1.21 illustrates a typical roadmap for the activities of project-by-project improvement teams. The tools shown here will be the subject of the improvement process explained in Chapter 4.

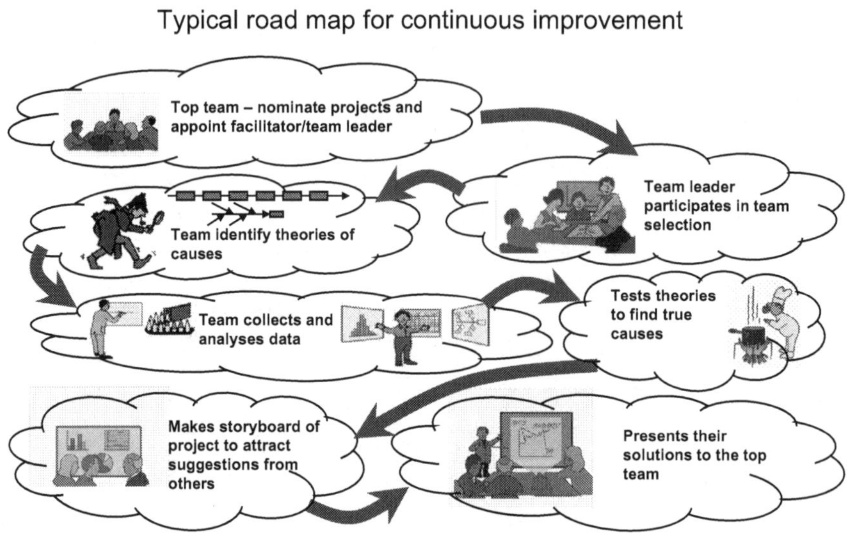

Typical road map for continuous improvement

Figure 1.21   Quality improvement organisation road map

Summary of learning

The two aspects of quality improvement are:

- The improvement of something that already exists but that requires improvement to performance
- Second, the prediction and elimination of potential problems that can arise following the development of new products and services to ensure that products and services come as close as possible to meeting the ultimate needs of the interested parties.

SECTION 8: 'EIGHT' MANAGEMENT PRINCIPLES

In Section 8 you will learn:

The importance of eight features that collectively make the biggest impact on the achievement of quality performance.

- Customer focus
- Leadership
- Involvement of people
- Process approach
- System approach to management

- Continual improvement
- Factual approach to decision-making
- Mutually beneficial supplier relationships.

> Note: Those who are familiar with the content of ISO 9000:2015 will say there are seven management principles.

The Standard ISO 9000:2000 stated that: 'Organisations depend on their customers and therefore should understand current and future customer needs, should meet customer requirements and strive to exceed customer expectations'.

The current version of the Standard ISO 9000:2015 has removed the principle covering 'systems' because it implies that the organisation is developing or improving a system. That is OK if you are reading it in the Standard whilst in the process of implementing it, but we have no idea whether those reading this book are doing that, so we have re-included it hence we are saying there are eight principles and not seven as stated in the Standard. We think that to have taken it out is confusing. Of course, an organisation must have a system of some sort, even it happens to be informal. So, we include it here as we believe that it is still relevant.

Principle 1: customer focus

This is true for existing products, but we think the situation is more complex when it comes to innovation. In this case, the customer does not know he/she has a need because the product does not exist. Nobody had a need for steam trains, television, electric lights, telephones let alone smart phones, so their appearance was not driven by need. In fact, even after the invention nobody complained about poor quality because they were only too happy to be able to use the service. Quality in this case is being the first into the market with a new and exciting product.

Principle 2: leadership

Leaders establish unity of purpose and the direction of the organisation. They should create and maintain the internal environment in which people can become fully involved in achieving the organisation's objectives.

We agree with all of that, but we think it should say more. It should also say that a good leader build the capabilities of people and builds teams of mutually supporting members. A good leader is trusted and respected by the team and by those with whom the team must interact with. Empathy is important in teamwork and a good leader is also a good listener, cares for the team members and their needs.

The ISO Standard states:

People at all levels are the essence of an organisation and their full involvement enables their abilities to be used for the organisation's benefit.

Principle 3: involvement of people

The Standard provides a list of potential benefits that a quality approach can achieve for an organisation, all of which are possible. Unfortunately, this principle, in our opinion, is more or less useless without any suggestion of 'how to'. We believe that its achievement requires the support of participative approaches such as quality circles or other forms of self-directing, or self-managing approaches in order to be effective. To be successful, an organisation needs to develop a culture in which everyone from the top to the bottom is involved in making their organisation the best in its business.

Principle 4: process approach

The Standard states:

> A desired result is achieved more efficiently when activities and related resources are managed as a process.'

This begs the question 'is there any other way of managing activities that have inputs and outputs?' Again, the lack of any 'how to' is a problem and needs to be addressed. Consideration should be given to the possible use of what Dr Juran used to refer to as the Quality Trilogy. In this concept, every person in the operations process is the customer of the former and a supplier to the next. Consequently, they are simultaneously, customer, producer and supplier to the previous or next person in the supply chain. Does our supplier meet our needs? Do we meet the needs of our customer? Have we ever asked?

Principle 5: system approach to management

The year 2000 version of the Standard stated:

> Identifying, understanding and managing interrelated processes as a system contributes to the organisation's effectiveness and efficiency in achieving its objectives.

This principle does not appear in the 2015 version, but we say:

The key benefits listed for this principle include: 'Continually improving the system through measurement and evaluation.' It is hard to see how 'measuring and evaluating' something will by itself improve it. All inspection can do by detecting a bad product is to remove it from the process. Since inspectors are only human, the likelihood is that they will miss many of the bad products (in most cases a minimum of 10% and generally many more) and quite frequently reject some of the good ones. All inspection can do to a good product is to add cost to it. It is better to make it good in the first place and reduce the need for inspection.

Principle 6: continual improvement

The Standard states:

> Continual improvement of the organisation's overall performance should be a permanent objective of the organisation.

As stated, this is a trite statement. Again, the Standard is silent on the 'how to'. In Chapter 4 we will go into a lot of detail as to how this can be achieved. We can always improve things by making observations but in a competitive world it is the rate of improvement that counts. If our competitor is improving at a faster rate than us then we need a structured and disciplined approach just as the British Olympic cycling team did in Beijing, London and Rio.

Principle 7: factual approach to decision-making

The Standard states:

Effective decisions are based on the analysis of data and information.

Since decision-making based on facts is the essence of total quality and is fundamental to the process of knowing what our actual target performance is, we need to know what is our actual performance and what is the gap between our actual performance and where we want to be.

Principle 8: mutually beneficial supplier relationships

The Standard states:

An organisation and its suppliers are interdependent, and a mutually beneficial relationship enhances the ability of both to create value.

This principle should also say that a healthy organisation cannot abdicate its responsibility for the quality of its products or services to a supplier. The ultimate responsibility always lies with the final producer whose product or service it is.

The supplier should be an extension of its own processes and control should always be in the hands of the outsourcing company. This topic is covered in depth in Chapter 7.

Pause for thought

Can you identify the eight management principles in your organisation? Check each one and ask yourself 'do you think they are adequate'? What changes might you make? Note them down.

Summary of learning

The eight principles of quality management were originally defined in ISO 9000:2005, Quality management systems – Fundamentals and vocabulary, and in ISO 9004:2009, 'Managing for the sustained success of an organisation'.

The quality management principles can be used as a foundation to guide an organisation's performance improvement. They were developed and updated by international appointees to ISO/TC 176, which is responsible for developing and maintaining ISO's quality management standards.

SECTION 9: PDCA

In Section 9 you will learn:

- PDCA stands for PLAN, DO, CHECK, ACT (or ACTION).
- How PDCA is incorporated in extreme work cultures and can be used to demonstrate the difference between the different management approaches.
- When is a problem 'worker controllable'?
- You will later find that PDCA turns up in later chapters as well.

The so-called PDCA cycle is absolutely fundamental to the creation of a well-designed quality programme because it is present in some form in everything that we do. PDCA stands for PLAN, DO, CHECK, ACT (or ACTION). (Note: Some today say PLAN, DO 'STUDY', ACT but we think this is pure semantics – the effect is the same.)

For example. I feel like a cup of tea. So, I create a PLAN. What sort of tea do I want? English breakfast, loose tea or a teabag? Do I want sugar and/or milk? Only milk. Do I want it weak or strong? Medium. Where shall I make it? Where are the utensils, etc.?

Now comes the DO. According to my PLAN, I put the kettle on, get a cup and saucer from the cupboard, a spoon from the drawer and find the tea bag in another cupboard. I put the teabag in the cup. When the kettle boils, I pour the water into the cup. I squeeze the teabag against the side of the cup until the water is dark. I pour on the milk until the tea is the desired colour.

Now comes CHECK. Does it taste how I like? No, it is too strong.

The final stage ACT – add some more milk. I try it again. Is it OK now? Yes. OK, I make a note of the colour. I suppose that if I was fussy, I might even take a photograph of it so that I have a visual standard. I am waiting to meet the person who might do that! If it were a restaurant it could be a good idea. At what point does 'just right' become either too strong or too weak? It is like asking 'how many whiskers make a beard?' But, basically, we are standardising on the improvement, so it makes sense.

Later, we make another cup of tea. We follow all the same steps but this time add a bit more milk. Maybe we make a note of that. This way, if others make the tea, they can get the balance right.

This is the quality improvement! Note, each time we go around the cycle there is an opportunity for continual improvement. All work has these same four elements.

The following shows another use of the PDCA cycle to illustrate two extreme work cultures. At one extreme is the craftsmanship system. A craftsman by definition is responsible for the quality of his or her own work just as the person was making

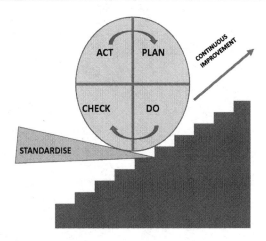

Figure 1.22   Holding the gains

a cup of tea earlier, so he/she has the whole PDCA cycle under their control. Any deficiencies could be regarded as 'operator controllable'.

At the other extreme is the so-called Taylor system or so-called 'scientific' management in which management manages and people do as they are told. Nobody asks them anything or involves them in anything. They just do their jobs according to instructions. So, they do not have control of the whole loop, so management must be implicated if deficiencies abound.

Later, we are going to show a different system that has all the advantages of both systems and none of their problems.

Figure 1.23 shows how the PDCA concept might be used to show the relationships between the workers and management under the Craftsmanship System. Note in this case the pivotal role of the foreman. He is the link between management and the workforce in this model.

It can be seen from the above figures that the PDCA model can show very clearly the differences between different management approaches. Figure 1.25 gives the workers the opportunity to be involved in improving the quality of their own work provided that they are properly trained to do so, and this is one of the main purposes of this book.

Pause for thought

PDCA in your workplace

For practice, try making a PDCA cycle for your job. Does the loop close? In other words, are all of the activities P, D, C and A under your or your team's control? If not, what are your conclusions? Does it mean that the operation is employee or management controllable?

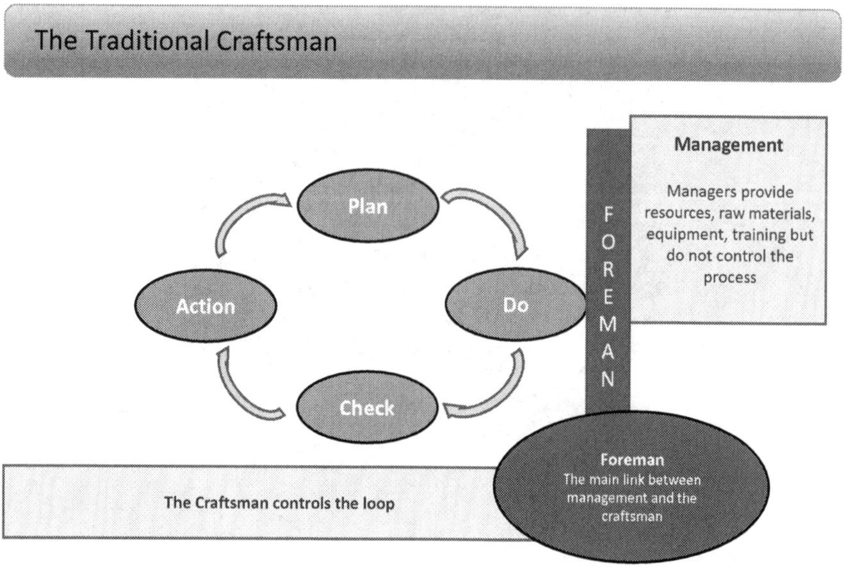

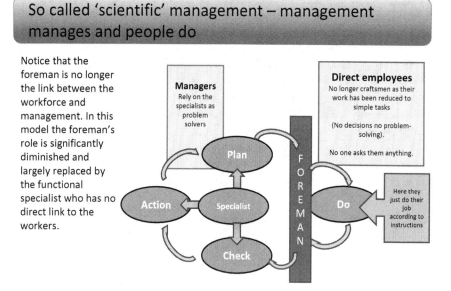

The Traditional Craftsman

Management

Managers provide resources, raw materials, equipment, training but do not control the process

Foreman
The main link between management and the craftsman

The Craftsman controls the loop

Figure 1.23   Craftmanship

So called 'scientific' management – management manages and people do

Notice that the foreman is no longer the link between the workforce and management. In this model the foreman's role is significantly diminished and largely replaced by the functional specialist who has no direct link to the workers.

Managers
Rely on the specialists as problem solvers

Direct employees
No longer craftsmen as their work has been reduced to simple tasks

(No decisions no problem-solving).

No one asks them anything.

Here they just do their job according to instructions

Figure 1.24   Scientific management

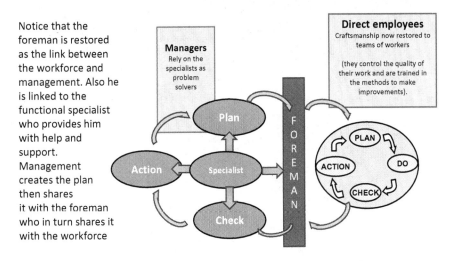

Professor Ishikawa's self control concept

Notice that the foreman is restored as the link between the workforce and management. Also he is linked to the functional specialist who provides him with help and support. Management creates the plan then shares it with the foreman who in turn shares it with the workforce

Figure 1.25   **Self control by workforce (quality circles)**

The principle of 'self-control'

The question often arises 'how many of the problems at work are management controllable and how many "worker controllable"'? This is important since when things go wrong it is always tempting to blame the workforce. Actually, it is rarely the case. For a problem to be 'worker controllable' it is necessary for the worker to have the four elements of the PDCA cycle under his/her control.

1   Knowing what is expected of him (the PLAN).
2   What is his actual performance? The DO.
3   The means to compare actual performance with expected performance (the CHECK).
4   The means of correction. (The ACT).

If any of these elements are missing the management job is incomplete, and any resulting deficiencies are management controllable not worker controllable.

Figure 1.26 shows yet another use of the PDCA model, which shows how the top management PLAN can effectively be cascaded down through the organisation through a succession of nested PDCA cycles. If any of these loops are broken it will result in deficiencies in the achievement of the PLAN. This can easily happen in cases where some managers choose to set their own agenda rather than participate in consensus-style management, which this is.

Participation

In the following version of the model the PLAN at the managerial level is MACRO plan in the sense that it represents the overall plan for the work to be carried out as

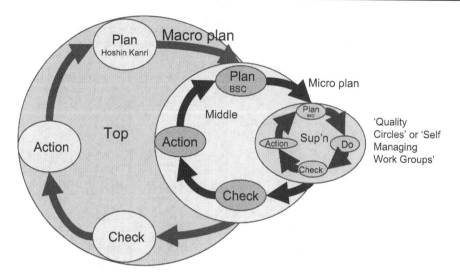

This development of Dr. Walter Shewhart's PDCA Cycle is just one of the concepts explained in the DHIQC Diploma in Quality Leadership. www.dhiqc.com

Figure 1.26   Nested 'do' loops

a whole including the daily management of the department. Ideally, this plan is shared with the supervisor or foreman who, in turn, shares it with the workgroup.

The circle or relevant work group with the participation of the supervisor/foreman will discuss and agree how to extract the elements in the PLAN that directly impact on them in order to create a 'micro plan', how this ~~the~~ micro plan will be achieved and what problems they may have to deal with, etc.

Pause for thought

Could you create a similar model for your organisation? Are all of the loops closed? If not, what might be the consequences? There is more on PDCA in later chapters.

Summary of learning

The so-called PDCA cycle is absolutely fundamental to the creation of a well-designed quality programme because it is present in everything that we do. PDCA stands for PLAN, DO, CHECK, ACT (or ACTION).

Each time we go around the cycle there is an opportunity for continual improvement.

The PDCA model can highlight the differences between different management approaches.

The following diagram shows how the complete management system of the organisation as a whole can be constructed to take advantage of the PDCA concept.

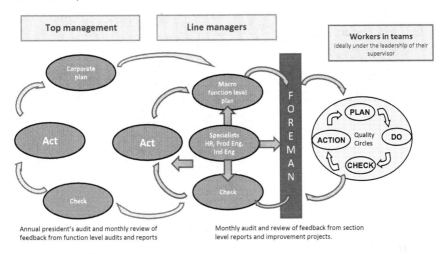

Annual president's audit and monthly review of feedback from function level audits and reports

Monthly audit and review of feedback from section level reports and improvement projects.

Figure 1.27   Macro view of participative management

SECTION 10: PROCESS DESIGN

In Section 10 we take a brief look at one of the most important aspects of successful business management, namely process design. This is not the same as 'product design', which is just one of the total sums of processes in an organisation.

The two main approaches are:

- Informal processes
- Formal or 'core' processes.

It is important because everything that happens in any organisation is an element in a larger process. Some processes are informal, not recorded anywhere and the tasks are not always carried out the same way. This is often the case in small organisations and is often referred to as 'shirtsleeve' management. Whether or not to record them will be to some extent dependent on the possible consequences if the process is subject to the likelihood of some dangerous situation if performed badly.

An example of 'shirtsleeve' management might be the cup of tea we made in an earlier section. It is unlikely that many people would record such a process unless it is something done repeatedly and, in a restaurant, where it is important that it is always done the same way in order to ensure predictable results

It might seem macho to manage the 'shirtsleeve' management way but generally with this informal type of management there is often a disaster just waiting to happen. The reason why formal systems are better is obvious. Generally, there is a best way of doing most things. If we know what it is, and we stick to it then we will get predictable

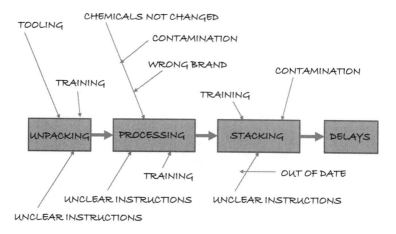

Figure 1.28   Process analysis

results. Also, if we know what we are doing and how we are doing it then it becomes possible to refine the process to get consistently better results.

Just watch any professional sports player and they are fastidious about their warm-ups, warm-downs, the training plan and everything. The gold medal winners record everything and if something goes wrong, they will quickly find out what it is and make some changes. Work is no different.

Every organisation comprises a multiplexity of interacting processes. These consist of what are referred to as 'core processes' and 'support processes'.

Core processes are those that are fundamental to the activities of the organisation, for example, the design, creation and delivery of the specific products or services for which it makes its living or its reason to be.

Figure 1.29   Medal winners celebrating

Figure 1.30   **Core processes**

However, these core processes require the support processes in order to be able to do this. For example, 'training' is a support process. Without proper training and skills development, it is unlikely that the core processes can deliver their outputs to a satisfactory level of quality.

We also need the finance process, scheduling, planning, maintenance, etc. These are all referred to as 'support' processes. Management of key business processes, support processes and the development of a complete management system are required to deliver customer and stakeholder needs.

In each of these processes, whether they be core or support, we can apply the SIPOC approach. SIPOC is an acronym for 'supply – input – process – output – customer', the triple role of process management.

There is an 'input' from its supplier. This input might in some instances be materials, but it can also include information in all its forms. With this information and or materials, operations take place, which is our 'process'. When completed, the result is passed as an 'output' to the customer of that activity or process.

Now this 'output' then becomes the 'input' for the next activity. Therefore the 'triple role' idea tells us that each person in a process is simultaneously a 'supplier' to the next upstream operation, it is a 'processor' and the role is to convert the goods and services provided by its supplier, into outputs for its customer. This completes the 'triple role'. The individual is therefore simultaneously a 'supplier', 'processor' and a 'customer'.

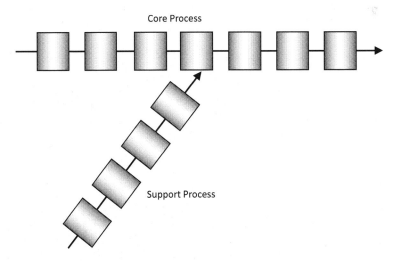

Figure 1.31   **Support processes**

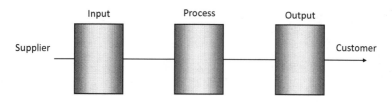

Figure 1.32   The SIPOC model

If we do not meet the needs of our customer there has to be a reason. Is it because we are not doing our job satisfactorily or is it because our suppliers are letting us down? Maybe both if we do not know the customer's needs.

Pause for thought

Quality in your workplace

- For your current operation make a list of your 'direct' suppliers.
- What is it that you require from them?
- What is it that they do that causes you problems?
- Do they know this? Have you told them?
- Now make a list of your current 'customers'.
- Do you provide them with what they need on time and every time?
- Have you ever asked them if they agree?
- Maybe there are some 'quality' problems here?

If the problem lies with our supplier, maybe it is because he does not know either. In this case he will keep on doing what he has always done until someone tells him something different. We look in more depth at supply chain management in a later chapter.

In a complex situation, therefore, it is not surprising that things do not always work out the way the end user would like. To deal with this we need first to define our processes, identify the internal inputs and outputs right through from the ultimate end user back through to the initial supplier. At each stage we need to check whether each of the inputs enables the next operation to satisfy the needs of its specific customer.

Summary of learning

Some processes are informal, not recorded anywhere and the tasks are not always carried out the same way. This is often the case in small organisations and is often referred to as 'shirtsleeve' management. Generally, there is a best way of doing most things. If we know what it is, and we stick to it then we will

---

get predictable results. Every organisation comprises a multiplexity of interacting processes. These consist of what are referred to as 'core processes' and 'support processes'.

SECTION 11: CUSTOMER SATISFACTION

In Section 11 you will learn:

- Measuring customer satisfaction
- The rules for good customer related data collection and analysis
- Benefits of customer data analysis.

It is not always appreciated that customer satisfaction is not the opposite of customer dissatisfaction. They might be at the opposite ends of a spectrum but separating them is 'customer indifference'.

Right from the word go, ultimately, the end user is the person we hope to satisfy, this is the person it was all for.

- Are they happy with us?
- How do we know?
- Should we measure the customer complaints?

The problem is, just because a customer does not complain does that mean he or she is happy? In short, we do not know and most probably have no idea. We also look at this again in more depth in a later chapter.

In order to find out what our customers think of our existing products, we need to find out through the following means:

- Questionnaires
- Shadow customers
- Interviews
- Non-customers
- Ex-customers
- Others in the trade.

Figure 1.33   Customer perceptions

We also need to study competitors' performance as well as our own. To do this we need to know:

- How to collect data
- Identify the attributes
- Design sampling
- Test the protocol
- Gather the data
- Analyse the data
- Use the data
- Improve the protocol.

To be successful the following are important:

- The programme must be fully supported by top management
- Address issues that are important to our customers
- Unbiased
- Statistically reliable
- Address strengths, weaknesses and competitive advantages
- Determine relative importance of key performance measures
- Comparable across business units
- Consistent over time
- Backed up with corrective action.

The objectives of the campaign will be to:

- Improve customer, client, or employee loyalty
- React quickly to changes in the market
- Identify and capitalise on opportunities
- Beat the competition
- Retain or gain market share
- Increase revenue
- Reduce costs.

Following the campaign, we need to feed the information to the people who might be able to do something in order to improve the results. Who might that be? Well, of course it depends upon the problems. In some situations, it might be the workforce in specific departments. In others, it could be within a specific function or levels of management or right at the top.

Pause for thought

Customer satisfaction in your workplace

Do you know who your organisation's key customers are? Do you know if they are satisfied? How do you know?

Decades ago, back in the 1950s, in fact, Dr Juran produced an estimate that showed that 20% of the problems were responsible for 80% of the cost of poor

---

quality (mostly resulting in 'customer dissatisfaction') and that this was management controllable and that 80% of the problems but only 20% of the cost was worker controllable. Could you produce a similar statistic for your organisation?

Summary of learning

It is important to develop the skills and processes to observe customers and to attempt to better understand their true needs. It is important to establish a baseline for your customer satisfaction measures. From simple surveys to interviews, the more you measure the more you understand the customer. However, it is an art and a science and there are important rules to follow if you're going to be successful.

CONGRATULATIONS!

You have completed 'Section 11', the last section of this chapter and so you have completed Chapter 1.

You have now already covered some of the fundamentals of quality and many of these will be treated in greater depth in subsequent chapters but hopefully you are beginning to see how a knowledge of these ideas can improve not only your life at work, in your family, with friends and in your own pastimes. After all, we all want to achieve the best quality of life, don't we?

2 *The role of people in quality*

AIM OF THIS CHAPTER

The human factor plays a prominent role in quality management. The aim of this unit is to show how quality depends essentially on people.

The learning outcomes of this chapter are:

- Understanding how people impact quality
- The ultimate aim is to create an organisation in which everyone is working towards making their organisation the best in its business
- Learn the six key activities that will make the move from left to right
- Understand what organisational values are
- Understanding the scope of organisational culture – Theory X management, Theory Y management, William Ouchi's Theory Z management and the Taylor system
- Good leadership – a natural quality or can it be learned?
- Using the PDCA cycle
- What are quality circles and the history of quality circles
- Who are the leading quality gurus?
- The role of steering committees
- Tuckman's theory of high performing teams, Belbin and the character types of winning teams
- Understanding communication and that one of the biggest problems in organisations is the lack of real communication between departments.

SECTION 1: HOW PEOPLE IMPACT QUALITY

What you will learn in this section:

- What 'quality' means to you
- Quality matters to people

- Working together
- How to 'sell' quality.

What 'quality' means to you

It is said that beauty is in the eye of the beholder. Well, it is just as true to say the same about quality. What might be considered to be 'quality' to one person might be considered to be trash to another.

Pause for thought

Think of something in your life where your opinion of quality is almost the opposite of that of your spouse, parents or kids!

The fact is that it is not only people who decide what quality is or is not, it is also people who create it. It might be created by individuals in isolation, for example, artists and composers, etc. Or it might be the result of very complex series of interactions between multiple groups of people whose combined efforts result in an outcome, for example, a rail service. This involves everyone from those responsible for laying and maintaining the track and signalling, etc., through the building of the trains, stations and ticketing, etc., to the staff who actually run the service. Any one or any group of these can at any time either positively or negatively impact the quality of the output.

It is a fact that apart from some sick people, nobody wakes up in the morning and thinks 'what can I screw up today?' We all go to work intending to do a quality job so if something goes wrong nobody wanted it to happen. So, there is no point in blaming each other. **We should attack problems not people.**

We get the best results by pulling together. This is the essence of a concept known as *Hoshin Kanri*, explained in more detail in Chapter 5, and which is a Japanese term meaning everyone in the organisation, all having the same vision and goals and working towards their collective achievement. It is well illustrated in the right-hand pyramid in Figure 2.1.

Organisations frequently look like the pyramid on the left in Figure 2.1. Think of the session that was recommended in the previous chapter with the two teams. Each part of the hierarchy is almost a separate organisation competing with all of the others. Frequently, the competition between departments is more intense than competition in the market place. The arrows depict internal conflict.

The picture on the right (Figure 2.1) shows an ideal organisation. Here, all the arrows are in alignment and show that the people in the various functions are all working towards the same objective. In this case, the only competition is in the market place with rival organisations. All internal competition is healthy, creative and not destructive.

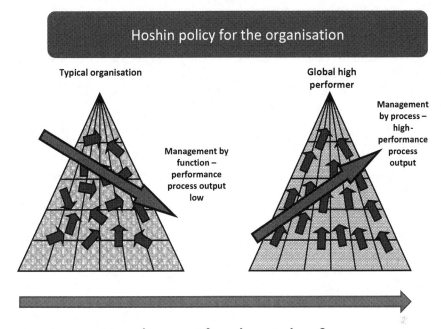

How do we get from here to here?

Figure 2.1   Management pyramids

Pause for thought

In your personal and business life think of situations where a team of what appeared to be very ordinary people achieved something collectively that nobody would have thought possible! More on this later.

So, how do we go from here to there? Well as you might imagine, if the organisation is firmly near to the left extreme, it is not easy to move to the right diagram in Figure 2.1, but the good thing is that every small increment along the way will produce measurable benefits.

Pause for thought

Actually, no organisation or group is likely to be right over on one or the other of the extremes. Most will be somewhere between the two. The important thing is, where is your organisation compared with its competitors and which is moving to the right the fastest? It might be a competing sports team not only your business rivals! Make a note of it!

The aim is to create an organisation in which everyone is working towards making their organisation the best in its business. The most important thing is for the chief executive to want to do it. No programme will work if the CEO is either lukewarm or hostile towards it. If you are not the CEO viewing this then the first and most important job is to try selling him/her the idea.

How do you sell it to a CEO or for that matter any of the top team? Remember, top management deals on the language of money. Just trying to convince them on the basis that it is a nice idea and that we can all dance off into the moonlight arm in arm singing songs, however appealing that might be, is unlikely to sell it to him/her.

The arguments they listen to are in the language of return on investment, profit and share of market, etc. Fortunately, we can use all these as will be seen in these chapters especially the one on cost of poor quality.

Another pause for thought

Can you think about how you could describe some of the quality problems in your work in terms of the language of money?

Not in any order, the key activities that will make the move from left to right in our two organisational extremes are:

- Set up multi-disciplinary project-by-project improvement teams
- Set up self-managing work groups within departments (quality circles or *Kaizen* activities)
- Create a top team steering committee
- Run the process down the supply chain
- Establish clear values for the organisation
- Adopt a *Hoshin Kanri*-style management system.

Summary of learning

In this section you will have learned:

- What we have learned in this section
- What 'quality' means to you (have you sorted that yet?)
- Quality matters to people
- Working together
- How to 'sell' quality.

SECTION 2: ORGANISATIONAL VALUES AND CULTURE

What you will learn in this section:

- What are 'organisational values'?
- Personal values
- Organisational values
- Quality as a value.

In the previous section we saw what it is that we want to achieve. In this section we will begin to look at the steps along the way towards how it might be achieved. The first step is to establish clear supportive organisational values that apply to all and to recognise the cultural issues that need to be addressed or reaffirmed.

We have a simple definition of a value: **something that is important to you.**

Value statements are often referred to as 'guiding principles' and can mean different things to different people depending on who writes them. They are only true values if regardless of circumstances they are bedrock and will never be changed due to expediency. If they are, they are just nice-sounding statements. Both companies and individuals can have value statements. A value statement tells the customer and the employee where the company stands and what the company believes in and what they can be trusted for. Sometimes they are posted on the wall in the reception area for all to see. This, also, sadly includes the 'nice-sounding statements'.

Pause for thought

As an individual, when you are living your values, it is when you are most fulfilled with life. For example, you could value spending time with your family, or being out with nature.

Values vary considerably from one person to another. Some may be handed down through generations, some may be learned from our parents and family, others from our communities and all may be modified by our own self will, perceptions and the impact of the values of others around us. Make a list of some values that you are familiar with in your family and social life as well as at work.

Figure 2.2   **Value statements**

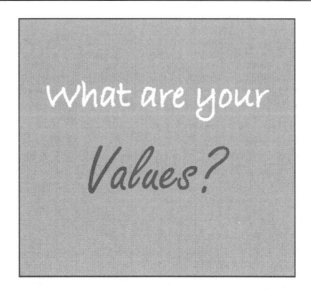

Figure 2.3   Choosing values

Organisational values

Organisationally, in theory, in a cooperative or perfect partnership, the values of the organisation should reflect the collective values of the population of that group. Some members may be more persuasive than others, in which case their values might dominate. In a large corporate organisation, the same is true but in this case it is likely to be the values of those who dictate policy, which appear to the market at large.

Values are important if an organisation intends to gain a reputation for quality especially its ultimate reputation in its market place.

Values matter. They form the very foundation of a business. They don't take long to define and articulate. Once you have them in place, they help guide the business forward. Honesty, or lack of it, is a rapid differentiator in terms of quality reputation and increasingly impacts such issues as environmental impact.

Quality as a value

Now, to a large extent our approach to quality is derived from our values. But it is also influenced by the work culture of the organisation. It is a fact that the values that are dominant on the inside of an organisation will be reflected in its behaviour outside.

If an organisation does not respect its employees and treats them badly how can it expect those same employees to behave any differently themselves when dealing with the company's clients? Theoretically, we would expect everyone to be in favour of 'quality'. Most people would like to see 'better quality'. Nobody enjoys the experience of being impacted by a poor-quality product or service. We feel shocked and cheated whenever that occurs.

Pause for thought

Reflect on your own, your group's and your organisation's values.
  Write down your thoughts.

Summary of learning

In this section you will have learned:

- What are 'organisational values'?
- Personal values
- Organisational values
- Quality as a value.

SECTION 3: ORGANISATIONAL CULTURE AND LEADERSHIP

In this longer section you will learn:

- Scope of organisational culture
- Theory X management
- Theory Y management
- Choice of culture
- Motivation
- William Ouchi's Theory Z management
- Empowerment
- The Taylor system.

The focus here is on organisational culture, which we are defining as 'corporate culture' and describes and governs the ways a company's owners and employees collectively think, feel and act. Organisational culture and leadership are elements in a company that work in conjunction with one another toward organisational success. Both culture and leadership influence how the company will function and what will be achieved. Either culture will determine how leadership functions, or leadership will transform the organisational culture so that the culture supports the organisational values.

There are two extremes of corporate culture. At one extreme, an organisation might be very autocratic with just the CEO and possibly his immediate direct reports making all of the decisions on behalf of everyone and using fear as the means to achieve its goals. At the other extreme, there are some very participative organisations that manage with a consensus style and attempting to involve everyone and to use their skill and creativity to achieve results.

An American behavioural scientist, Douglas McGregor, in the early part of the twentieth century, defined these categories as being **Theory X** (autocratic) on the one hand, and **Theory Y** (participative) on the other (*The Human Side of Enterprise*, 1960).

The differences in these two styles have led to the creation of two distinctly different work cultures. Paradoxically, both work and both can, and do, exist in any society. Whilst these are extremes, there is a tendency, at least in the West, for organisations to be biased towards the Theory X side of average.

Theory X: authoritarian management

This type of manager tends to believe:

- The average person dislikes work and will avoid it if he/she can
- Therefore, most people must be forced with the threat of punishment to work towards organisational objectives
- The average person prefers to be directed; to avoid responsibility; is relatively unambitious and wants security above all else.

Theory Y: participative management

This type of manager tends to believe:

- Work is as natural as breathing and sleeping
- People will apply self-control and self-direction in the pursuit of organisational objectives, without external control or the threat of punishment
- Commitment to objectives is a function of rewards associated with their achievement
- People usually accept and often seek responsibility
- The capacity to use a high degree of imagination, ingenuity and creativity in solving organisational problems is widely, not narrowly, distributed in the population.

In industry, the intellectual potential of the average person is only partly utilised. The most striking difference between the two approaches is the belief by those biased towards Theory X that people can be forced to conform to specification requirements through auditing and the adoption of a 'blame culture' – 'whose fault is it' when something has gone wrong?

This approach to management appears to work well where there is what can be referred to as a 'Taylor system' (or, alternatively, 'scientific management') approach. It is called the Taylor system after Frederick W. Taylor, an American mechanical engineer, advocated in the 1920s that 'Management managed and People just "did their jobs according to instructions" to improve "efficiency"'.

Although this approach might have been endemic 100 or so years ago and has been widely criticised for more than 50 years, the system has been proved persistent and is still not far from the surface in many Western organisations, the reason being that it is self-sustaining. Systems that are heavily dependent on regulation and standardisation tend to operate according to this approach as its exponents erroneously believe that the results are more predictable. For example, the Theory X-style manager who rules with a rod of iron gets results while he is looking. However, as soon as he turns his back, performance drops off. This is noticed and so his beliefs are reinforced.

However, this is also the case for Theory Y. The Theory Y manager is usually liked and respected. People like working for him/her. As a consequence, if the Theory Y manager is away from the job, the people will continue to perform at their best. This also reinforces the belief system that people want to participate, again, making the approach self-sustaining.

It can be seen, therefore, that these belief systems are not easy to dislodge and it is difficult to change from one to the other but it can be done. The payoff is that Theory Y produces consistently significantly better results than Theory X. So how can we change?

What has proved to be the best way is not to try to do too much all at once. A small-scale try-out is best. Start where the grass is greenest. Some Theory X managers will be very wary of Theory Y. They are convinced that Theory X has worked for them and they will be unwilling to try anything different resulting in their loss of control. However, other managers or supervisors will be more courageous. Perhaps they lean in the direction of Theory Y anyway. Once they have got started and the results become apparent some more Theory X managers will want to join in and so it spreads not by force but voluntarily, which is the best way.

It is also worth reading Frederick Herzberg's classic book *The Motivation to Work*, written in 1959! It's a pity this was not available at the time of Frederick W. Taylor! It might seem old today, but the message is just as relevant as ever it was.

Managers often ask: 'How can I motivate my staff?' The answer is that you cannot. Motivation is under their control not yours. You cannot actually motivate them, but you can stop **demotivating** them!

There are three factors that influence work:

- Motivation
- Work environment
- Management behaviour.

You can change the work environment and you can change management behaviour but 'motivation' is in the control of the workers. Create the right environment and the workforce might get motivated.

To create a positive environment, it is necessary to use empathy.

- How would I like it if I was treated like that?
- How would I feel if?
- Nobody asked me anything
- Nobody told me anything other than feeding me with unexplained instructions
- Nobody has taken an interest in my development
- Nobody says thank you for a job well done
- I see no prospects for any change in the future.

Pause for thought

Ask yourself. How would I feel if ... ?

- I went to work with the feeling 'I just hang my brains on the gate when I come to work in the morning and pick them up again at night'.

- Nobody asks me anything, nobody involves me in anything. I just have to do my job according to instructions day in and day out.
- How would I feel if I got that sick feeling every Sunday night, God it is work again tomorrow and I have to grind through a whole week of it to bring home some pay to feed my family?
- How motivated would you feel? Maybe you do feel like that! It does not have to be. Not for anybody.

On a scale of 1 to 10 with 10 being Theory X, where would you put your organisation – honestly?

William Ouchi says that Japanese consensus management style is based on the assumptions that employees want to build cooperative relationships with their employers, peers and other employees in the firm.

For this, they require a high degree of support in the form of secure employment and facilities for development of multiple skills through training and job rotation, they value family life, culture and traditions, and social institutions as much as material success, they have a well-developed sense of dedication, moral obligations and self-discipline, and they can make collective decisions through consensus.

In Ouchi's book *Theory Z Management*, he compares organisations that are based on each of Theory X and Theory Y, their strengths and weaknesses and then postulates a Theory Z, which has the good from each and not the bad. Unfortunately, his book was published at about the same time as *In Search of Excellence* and was smothered by the publicity for the latter. Fortunately, we discovered it! It is well worth reading though sadly out of print, but second-hand copies can be obtained online.

We can call this approach 'empowerment'. In general, the world is slowly moving towards this approach and we will go into it in some detail. It is at the core of this chapter and, in fact, it underpins the philosophy responsible for the modern trends in all the other units as well. This is the main reason why we encourage our students to study this one before working on the others.

The empowerment approach as we know it today was first enunciated by the late Professor Ishikawa in Japan following World War II.

Each person is the expert in his or her own job. It is only by galvanising the knowledge, skill and creativity of all the people in an organisation that we can strive to become the best in our businesses.

(Professor Ishikawa 1979, speaking at a conference)

We will call it Japanese TQM (total quality management) but it is also fundamental to *Hoshin Kanri*, which is the umbrella over TQM.

The term TQM is used extensively in the West as well, but the meaning is founded on compliance-based systems not collaborative ones.

We are going to win, and the industrial West is going to lose out because the reasons for your failure are within yourselves.

Your firms are built on the 'Taylor' model. Even worse, so are your heads. With your bosses doing the thinking while the workers wield the screwdrivers, you're convinced deep down that this is the right way to run a business. For you, the essence of management is getting the ideas out of the heads of the bosses and into the hands of labour.

We are beyond the 'Taylor' Model. Business, we know, is now so complex and difficult, the survival of firms so hazardous in an environment increasingly unpredictable, competitive and fraught with danger, that their continued existence depends on the day to day mobilisation of every ounce of intelligence!

(Konosuke Matushita 1979)

The so-called Taylor system

As stated earlier, the term is named after the American management expert in the early part of the twentieth century named **Fredrick W. Taylor**. In some ways it is unfair to name the concept after him because its origins date centuries before his time.

Crudely speaking, the so-called Taylor system can be described as a management system in which **managers manage and workers do**.

- Work is broken down into its individual elements
- Workers perform the same tasks repeatedly whilst management controls the process
- Workers were timed for each element of their work in order to set piece work rates – 'Time and Motion'
- Typically, common in high volume manufacturing.

Taylor's work had a profound effect on the nature of work organisation. It might not have been Taylor's intention but whilst it had a major impact on productivity, low unit cost, interchange ability of parts to mention just a few, the negative social consequences were equally profound as Karl Marx had prophesied before Taylor was born and Marx even founded communism or his perception of it.

By the continual deskilling of work, it will eventually become totally meaningless to the worker. As it continues, the stomachs of the bosses will become fatter whilst the arms of the workers become thinner. Under such a regime the only way that the workers can obtain their needs is through conflict.

(Karl Marx 1867)

Later, Professor Ishikawa had a different view, as we will see later. Karl Marx's view is not exactly what happened, but he was close. Work itself certainly did become meaningless, not just for the workers in factories but eventually across the whole scope of business life. Today, at the beginning of the twenty-first century, we have moved a long way from that concept but probably by no means far enough.

Socrates and Plato referred to the Division of Labour conceptually as did Adam Smith in his famous book *The Wealth of Nations* and the original concept now referred to by many as Taylorism but work place organisation is still a very long way from being ideal.

The first well-known and highly dramatised application of Taylor's thinking was at the Ford Motor company during the Great Depression. Henry Ford challenged the traditional method by which automobiles were produced. Instead of having the vehicle under construction being built by a team of workers on a wooden platform (which is how they were built at the time and in the UK even into the late 1950s), where the workers came to the vehicle, he instead created what we now know as the production line.

From a production point of view, the effect was amazing. Instead of making a few vehicles in a week, they were able to produce hundreds in a single day (93 minutes per unit!). The price of the finished product dropped. By the 1920s, the price had fallen to $290 (equivalent to $3258 today) when competing cars cost between $2000 and $3000 because of Ford's increasing efficiencies of assembly line technique and volume.

Even in the Great Depression this meant that over 50% of the population of the United States could afford a car, which cost just four months' salary for a Ford car worker. Ford quickly became one of the highest-volume producers in the USA, but it was not before long that Ford's competitors copied his methods.

Before World War II, this had become the standard method of production in the USA for all products. Later, the concept spread to offices, hospitals and, in fact, everywhere that people worked but, interestingly, it only reached UK car manufacturing in the late 1950s.

Life before Taylorism was referred to as the 'craftsmanship era.' Contrast the concept of a craftsman with that of the division of labour approach. The main problem is that while craftsmanship leads to much better quality in terms of functionality and appearance, harmony in the workplace, etc., Taylorism is vastly superior in terms of efficiency, volume of production and cost. Taylorism also has merits in terms of interchangeability of parts.

In most cases, what appears an advantage for one of these work systems becomes a disadvantage in the other. For example, 'motivated and involved', which is a positive feature in a craftsmanship regime, becomes 'unmotivated and not involved' in a Taylorised environment. Indeed, 'how to motivate the workers' has been a perennial problem ever since Taylorism was first introduced, hence the work of such pioneers as Herzberg and Maslow.

Adam Smith's case example – the pin factory

One of the first people to describe the division of labour was Adam Smith in his famous (1776) example of a pin factory. Smith also first recognised how the output could be increased using division of labour. Previously, in a society where production was dominated by handcrafted goods, one man would perform all the activities required during the production process, while Smith described how the work was divided into a set of simple tasks, which would be performed by specialised workers.

The result of labour division in Smith's example resulted in productivity increasing by 24,000% (sic), i.e. the same number of workers made 240 times as many pins as they had been producing before the introduction of labour division. The methods used in the pin factory are far superior in terms of volume production but how about quality both in terms of quality at life at work and quality of the finished product?

Pause for thought

How much evidence is there in your organisation of any remaining aspects of Taylorism. Honestly, do you think that your people are fully empowered? Including yourself, how many get that sinking feeling on a Sunday evening – Hell, it is work again tomorrow? They must do it week in and week out just to feed themselves and the family. It does not have to be like this!

How many on your payroll possibly left school, found a job and since then nobody has given a thought to how whatever talents they might have could be used and developed? Does this seem like a waste to your organisation and a no-hope future for the individuals?

Summary of learning

In this section you will have learned:

- Scope of organisational culture
- Theory X management
- Theory Y management
- Choice of culture
- Motivation
- William Ouchi's Theory Z management
- Empowerment
- Taylorism.

SECTION 4: LEADERSHIP

What you will learn in this section:

- What is a leader?
- Types of leadership
- Good leaders.

Learning leadership

Who are we talking about when we say 'leader'? Well, surprisingly it can be you regardless of who you are. It is not just the boss, the manager or supervisor because all of us find ourselves in the role of leader at some points in our lives. We might not always be regarded as 'leaders' in our work but outside of work, we assume that role either voluntarily or otherwise probably over and over without even thinking about it.

In work or outside of work, it makes no difference, the role is the same and we can deal with it in much the same way whether it is a formal or an informal role. Some examples of non-work-related leadership roles might be: scout or girl guide master/leader, protest march, mother or father (we lead the family), local kids' football coach, organising a party, wedding, funeral, christening, even planning for the family to come down for Christmas or other festive occasion. Even organising a raffle with a few helpers involved makes us a leader at least for a brief period.

In all cases, whether in work or outside, we can treat the role of leader in the same way as in the other but there are extremes of leadership style. At one extreme we might think, great, now I have power over all those people. I can boss them around and they must do what I say. If you think like this, though, and it is a voluntary role you might soon find you have a team of just one! You!

Or, alternatively, you might think, I am now in a position of trust. I am now responsible for the enjoyment of life of all those people who either volunteered or were assigned to join my team. They are looking up to me, have expectations of me and I do not want to disappoint either them or those who have expectations. The team was formed for a purpose. Let us make sure that everyone knows that purpose and that we can agree their role in achieving it.

All we can say to those who are attracted to the former of these options is to re-examine their motives. Sooner or later, in many cases sooner, they will fail and fail dismally. They will end up with few friends and many enemies. Nobody likes either power-hungry autocrats or control freaks. They may get their way for a time but sooner or later they will trip up or be tripped up! And nobody will want to help them.

On the other hand, for the participative leader it is a very different story. Not only does the participative leader really care about the members of the team, he/she instinctively realises that the group is brimming with talent. There are all sorts of skills that they are only too pleased to put at the disposal of the team. You do not even have to have prima donnas.

Pause for thought

Think about some situations where you have been the leader either in work or in your private life. Ask yourself, what sort of leader was/am I? Would others agree with my opinion. Is it OK to ask? Are you making notes? Keep the pad to yourself!

A winning team can consist of very average people but when they come to work together, they are invincible. Just look at Leicester City Football Club in the 2015/16 season. They were almost undefeated throughout the year and there was hardly a star amongst them. So far, they have not repeated it in following seasons but maybe it is because they have not managed to find the sparkle that set them alight the year in that memorable year. It is the good leader's job to find that sparkle.

We are concerned with the following kinds of leadership in this section.

1. Formal imposed leadership such as that to be provided by supervision, line management and corporate management, and
2. Informal consensual voluntary leadership such as when a member of a group voluntarily takes charge of a situation and the group agrees or when the

group choose a leader from amongst themselves to achieve some self-selected objective. This might be for the organisation of some function or other or maybe to deal with a grievance of some sort.

Somewhere between the two, there is a formal type of leadership selected by the workforce and which may or may not be acceptable to management as in the case of a trade union representative.

Good leaders

Leaders create a vision. They have a picture of the future and a vision of success. What does it look like? What does it feel like? Can we transmit the positive feelings to the others on the team? Can it be described so others can see it and feel it, get passionate about it? The vision may have no precise goals or numbers, but it does describe a future reality. It may be difficult to achieve, but not impossible. Often 'stretched goals' are more stimulating than easy ones but they must be believable.

Leaders build alignment. They energise people to commit to the vision and articulate how to achieve it. Leaders may achieve alignment through a variety of different ways such as rousing speeches, charisma, personal loyalty and rational arguments.

Leaders deploy. They take finite resources and determine how to achieve their visions. These deployment decisions form the strategic roadmap.

There are other merits too. Imagine that you have just left school and are starting work as an apprentice. In some cases, your parents might even have to pay for the privilege of your learning your craft from a highly skilled man! This is how it was for most apprentices 100 years ago and more. Who would your heroes be in your world of work? Generally, it will be those who know the most and can do more than you. Who is the ultimate hero?

In most cases it will be the acknowledged leading craftsman. Who is that? Usually, but not always, it will be the foreman or supervisor. That is why they were given that position. Part of their job is to help others achieve their level of skill and the boss or owner will depend heavily on them for the quality of the products produced. In this case they may sometimes be an informal leader. He may not even know that you have chosen him as your leader. We sometimes refer to such people as our 'heroes' or 'gurus'. If they do know it then they have a responsibility.

Of course, it is also to make sure that the work gets done efficiently and effectively and to communicate with the boss. He would, in fact, be the key link in the relationship and communications chain between the owners and the workforce, in both directions (up and down).

One key question is: is the ability to lead teachable or is it a capability that some people have quite naturally? Some people seem to have a natural ability to lead even with no formal training whatsoever. It just seems to come naturally. But even in this case, they can learn to lead more effectively. Also, it is certainly true that bad leaders can be dramatically improved if they are prepared to acknowledge their weaknesses and work on them

This is the last point in this short section but do not think that it must be the end of your learning about leadership. Of all the topics in this book, leadership is one topic where we can never cease to find new opportunities to learn how to become a better and better leader. Leadership and teamwork are interrelated topics. Of course, there may in certain situations be teams comprising a leader with a team of just one person. Often, in this case the leader is referred to as a 'coach' or sometimes 'mentor'. But in all cases, we do not have to go to the classroom as we can learn a huge amount in the University of Life.

What do we mean by that? Well, just in this section we mentioned Leicester City Football Club's 2015/16 season for their stunning success resulting from brilliant leadership and stunning teamwork but there are other examples all around us such as the GB cycling team. Before Sir Dave Brelsford was appointed as head, the GB team was nothing special. Yes, it had a few successes but nothing spectacular then look what happened in the Beijing Olympics, the London Olympics the Tour de France and Rio. The cyclists were the same, but the results were very different!

All of us can do that both in our private lives and in our work. The secret? Working together under good and trusted leadership.

Pause for thought

Think back through your life. Was there a time where you found yourself in a team under an inspiring leader? What was the situation? What was the task? What did it feel like being on the team? What do you think of the leader at the time? How did you feel when it was all over?

What can you learn from it and make happen on purpose what at that time might possibly have happened by luck?

Summary of learning

In this section you will have learned:

- What is a leader?
- Types of leadership
- Good leaders
- Learning leadership.

SECTION 5: PDCA

In this section you will learn:

- What is PDCA?
- Using the PDCA cycle
- Using the PDCA for process improvement
- PDCA – open loop systems
- PDCA illustrations of use in management systems, craftsmanship, Taylorism, QC circles, management or worker-solvable problems.

We took a brief look at PDCA in Chapter 1 but let us now look in bit more depth.

Actually, it is very ordinary really. Many people in the quality profession think it is relatively new if 35 years can be counted as being new but, in reality, the concept was known at least 200 years ago but not by those four words. They did not appear until the 1920s.

PDCA, as it is often known, is, in fact, very important to us and unconsciously it is something we use mostly without realising it every time we do almost anything, even brewing a cup of tea. The acronym stands for PLAN, DO, CHECK, ACT. We will explain it in minute. It is a simple concept, but it is important because, by understanding it, it helps us to investigate processes and why they either work or do not. If a process does not work as well as we would like it is possible that there is a flaw in the content of the PDCA cycle as we will see. Before we start to analyse it let us see how it does fit making a cup of tea.

PLAN

Let us see how it works by looking at the process of making a cup of tea. Before we do anything, we consciously, or often unconsciously, make a PLAN. In the plan the thoughts that whiz at lightning speed through our minds are: is the workspace clear, is there a kettle and socket, do we have any tea, milk and sugar, etc., cups saucers, spoons, supply of water. If we have guests, we may ask them how they like their tea – strong/weak? With or without milk or sugar, etc.? Teabags in cups or teapot? OK, bags in cups.

DO

With the plan in place. We make the tea. We put the kettle on, wait for it to boil, get cups, put teabags in, pour on the boiling water. Get the milk from the fridge, pour onto the water. Put sugar in for those that want it. Stir the tea.

CHECK

The tea is now brewed. Is it the right strength? Taste a sample or check by sight. Put cups on a tray and then serve. There is a comment from a guest that the process took a long time.

ACT

Why did the process take too long? What can we do to speed it up? An idea. What if we get the milk from the fridge, lay out the cups, spoons and get the sugar during the boiling time. This is wasted time! We add this to the plan for the future. So, in just one iteration of the PDCA cycle we have made a quality improvement that will occur every time we make a cup of tea in the future. This was a simple example and PDCA can be used in quite complex situations, but the principle remains the same

For example, we can summarise the contrasting leadership styles of the previous section and their impact on organisation and quality performance using the PDCA cycle. It is a very simple but powerful concept. We said previously, the PDCA cycle predated even Dr Shewhart who introduced it into the field of 'quality control' in the 1920s. Prior to that it was known as the **closed feedback loop as it applied to servo mechanisms**. An example would be the fully closed loop-controlled air conditioning system. This system would require both the means to raise and lower the temperature of the location in which it is situated.

Here are two illustrations of the use of the PDCA cycle using a central heating system.

1. Open-loop system.

It is referred to as an open loop system because whilst the thermostat can tell the boiler to heat up the water and switch on the pump, if the outside temperature goes above the temperature set on the thermostat it can switch off the boiler but there is no way it can cool down the room.

Here is a second illustration of the use of the PDCA cycle using a central heating system.

2. Closed-loop system.

It is referred to as a closed-loop system because the stat can now tell the AC unit either to heat up if the temperature is below that of the thermostat or cool it down if it is above, so it gives full control and is known as a closed-loop system, which is ideal.

We saw from the 'cup of tea' example in Chapter 1, that the whole process was in the hands of the person making the tea. As we saw later, this was the essence of 'craftsmanship'. If, therefore, the tea had not tasted so good, the cups were dirty, then the problem was caused by the teamaker. It was completely an operator-controllable process.

Pause for thought

Study the steps in the process of your own job. Identify which elements are the PLAN, DO, CHECK and ACT features.

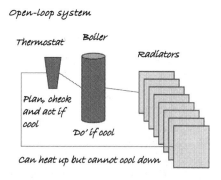

Figure 2.4   Open-loop system

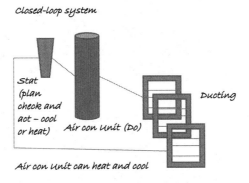

Closed-loop system

Stat
(plan
check and
act – cool
or heat)   Air con Unit (Do)   Ducting

Air con Unit can heat and cool

Figure 2.5   Closed-loop system

Whether or not an operation is either management solvable or operator controllable is an important principle and was brilliantly highlighted by Dr Juran. He stated that for an operation to be operator controllable, in addition to the DO element it required three key features.

1. The operator must know what is expected of him (PLAN).
2. Have the means to do good work (DO).
3. Have the means to check his work against the plan (CHECK).
4. Have the means to take remedial action (ACT).
5. Dr Juran stated that if any of these elements is not operator controllable then the management job is incomplete, and the resulting failures cannot be regarded as operator controllable. The following is a variation of an explanation of this concept by Dr Juran. It is one of the best illustrations of the distinction between operator and management controllable errors that we have seen and if you look carefully you can see the shadow of PDCA in the background.

Let us test this idea with an office operation, booking flights at a travel agency.

We have three operators who use a document to book the flights and there are multiple error types.

To do the study we create a matrix.

Let us just look at Operation 3.

We can see from this and the use of the PDCA cycle that it is not so simple as to say that problems are caused by the operators or the workforce. It is also not a solution to say 'motivate the workers' when the workers need an answer to the question 'What should I do differently from what I am doing now?' We can see that there are multiple types of what are often referred to as operator-controllable problems, some of which most definitely are not operator controllable as the management job is incomplete. So, if we are to solve these problems what can we do? Simply 'motivating the workers' is not the solution to their logical question 'What should I do different from what I am doing now?'

We can see from the PDCA that where problems really are in the hands of the workforce, such as was the case with the time it took to brew the tea, then the

Here is the matrix of error types/operator

Error type	A	B	C	D	E	F	Total
	Operator						
1	0	0	1	0	2	1	4
2	1	0	0	0	1	0	2
3	0	16	1	0	2	0	19
4	0	0	0	0	1	0	1
5	2	1	3	1	4	2	13
6	0	0	0	0	3	0	0
Misc	3	3	1	2	23	4	36
Total	6	20	6	3	36	7	80

Figure 2.6   Matrix of errors

We can see that operator B made 16 of the 19 errors. What about the rest of the work of operator B? We can see she is very good and makes very few errors. Clearly she is misinterpreting that instruction. We can solve that easily.

Error type	A	B	C	D	E	F	Total
	Operator						
1	0	0	1	0	2	1	4
2	1	0	0	0	1	0	2
3	0	16	1	0	2	0	19
4	0	0	0	0	1	0	1
5	2	1	3	1	4	2	13
6	0	0	0	0	3	0	0
Misc	3	3	1	2	23	4	36
Total	6	20	6	3	36	7	80

Figure 2.7   Operator errors

operators themselves are the best people to find a solution because it is only they who know exactly what they do. Even if management does study worker practices, that is no guarantee that it is the way things are always done especially if there is a Theory X management style!

If we want to close the loop on the PDCA then it can only be done with the full involvement of the workforce, which brings us to the next section on self-managing work groups or quality circles.

Notice for error type 5, all operators make that error rather uniformly. Experience indicates that when all operators make the same error it is not generally an operator controllable error. It requires management action to find the cause and implement a remedy. Clearly there are problems with the PLAN!

Error type	A	B	C	D	E	F	Total
				Operator			
1	0	0	1	0	2	1	4
2	1	0	0	0	1	0	2
3	0	16	1	0	2	0	19
4	0	0	0	0	1	0	1
5	2	1	3	1	4	2	13
6	0	0	0	0	3	0	0
Misc	3	3	1	2	23	4	36
Total	6	20	6	3	36	7	80

Figure 2.8   Collective errors

Now let's look at the biggest number of all the errors made by operator E. We could find a lot of causes but maybe it could be better to find a job more applicable to operator E's capabilities?

Error type	A	B	C	D	E	F	Total
				Operator			
1	0	0	1	0	2	1	4
2	1	0	0	0	1	0	2
3	0	16	1	0	2	0	19
4	0	0	0	0	1	0	1
5	2	1	3	1	4	2	13
6	0	0	0	0	3	0	0
Misc	3	3	1	2	23	4	36
Total	6	20	6	3	36	7	80

Figure 2.9   Personal errors

Pause for thought

Look at your and others' work. Is the loop closed? Are the specific operations fully operator controllable or is the managerial job incomplete? The Japanese solved this problem using quality circles and other forms of self-managing teams. What are your thoughts?

Summary of learning

In this section you will have learned:

- What is PDCA?
- Using the PDCA cycle
- Using the PDCA for process improvement
- PDCA – open-loop systems
- PDCA illustrations of use in management systems, craftsmanship, Taylorism, QC circles, management or worker-solvable problems.

SECTION 6: QUALITY CIRCLES

What you will learn in this section:

- Definition of quality circles
- History of quality circles
- The essence of quality circles
- Quality circles and the PDCA cycle
- Developments from quality circles
- Advantages of adopting quality circles
- Other types of performance improvement teams
- Storyboards
- Statistical process control
- Impact of quality circles on the West.

Quality circles (referred to in Japan as QC Circles) are small groups of workpeople who do similar related work and who voluntarily meet together in work time, on a regular basis and are trained to identify, analyse and solve work-related problems and to share the results in presentations to their supervisor. In some cases, the supervisor may also be a member of the group, in which case the presentation is made to the manager. *Gemba Kaizen* is an alternative name for the groups.

The concept was originated by Professor Ishikawa in Japan in 1962 following four years of development of the foremen and supervisors in factories outside of Japan. The idea grew directly from Dr Feigenbaum's observation that 'Quality is everybody's business' and this concept is also the foundation stone of this series of books.

Professor Ishikawa's suggestion that led to the creation of quality circles was made in a workshop magazine intended for the use of foremen and supervisors and was called 'Gemba to QC' (see Figure 1.3). The article was published in 1960 and again in 1962.

The Nippon Wireless and Telegraph company was the first company to make a start in 1962 and reported that it had set up nine of these teams and that they were incredibly successful.

By the end of that year, 35 other companies had followed suit, and all reported similar results. Labour discontent had all but disappeared, productivity had increased significantly and product quality was improving at a rapid rate.

By 1979 there were reported to be more than 1 million such teams operating in Japanese factories involving over 10 million Japanese workers. By this time, it had truly become accepted and institutionalised as 'the Japanese way of management'. Let us now consider this different approach to management with the aid of the PDCA cycle, which we studied in an earlier session.

In effect, when an organisation is managed in this way, all the advantages listed above for both craftsmanship and Taylorism can be realised. Actually, it goes much further than that. In Chapter 5 you will learn about project-by-project improvement, lean manufacture, and Six Sigma. These concepts all owe their success in Japan to the original development of QC circles.

The converse is also true. If QC circles had never got started, none of these concepts are likely to have emerged. Quality circles were originally called QC circles and are still known as QC circles throughout the Far East. The Americans dropped the word 'Control' but they are identical in every other respect.

In Figure 1.26 there are several things to notice. Firstly, the individual DO has been replaced by the group of workers as a whole. In effect, the management PLAN is communicated to the foremen. He in turn communicates it to his 'team'. The team are given time to consider the PLAN and together with the supervisor, create a mini PLAN of their own designed to meet the requirements of the management PLAN. They are trained in the tools of problem solving and allotted time each week to work on their PLAN. These teams of workers are generally called 'self-directing work teams' but the approach is also known as QC circles, *Kaizen*, simply 'small group activities', zero defects teams, discover loss groups, etc. and other names specific to the organisation that supports them. However, in Japan, regardless of the specific labels used by the different companies, the underlying philosophy is virtually identical.

Following the origination of this approach in Japan in the early 1960s it very soon became apparent that these 'teams' could be used as a vehicle to introduce an unlimited range of tools and techniques to the workforce and that is what happened. Concepts known in the West as 'lean', 'Six Sigma', etc. can all be traced back to their original roots in quality circles in the late 1960s and early 1970s.

1. QC circles are intended to strengthen the existing organisation, not create by-passes or alternatives.
2. QC circles will help to create a more co-operative work environment and build bridges between all institutionalised activities.
3. QC circles represent no threat to any part of the organisation including the trade unions.
4. Top management must be totally committed to every aspect of the concept if success is to be ensured.

5. Middle or departmental and functional management are the key factors in determining the success of circles within that segment of the operation.

6. The supervisor is usually the circle leader. Others may lead circles later when the programme is well established.

7. The support of trade union representatives is extremely valuable and is consistent with the objective of achieving greater involvement for their members and helping to create a better work environment. QC circle programmes should not cut across any legitimate union interest.

8. The specialists in an organisation are vital to QC circle projects, particularly when their special skills and knowledge are required.

9. Non-circle members should be given the opportunity to become involved wherever possible.

10. The steering committee provides a cross-functional supportive framework for circle programme development

11. The facilitator is the focal point of the programme and in most cases should be a full-time appointment. Refer to the final chapter for details.

Self-directing work teams are not the only types of improvement team and many have their roots other than in Japan. For example, because of Dr Juran's 'breakthrough' process (covered in Chapter 4), many organisations around the world have implemented a team approach to business improvement.

Many of these could be described as QC circles, which are also known as 'self-directing work teams' or *Kaizen*, and these are voluntary. But there are also cross-functional teams, blitz teams and Six Sigma teams.

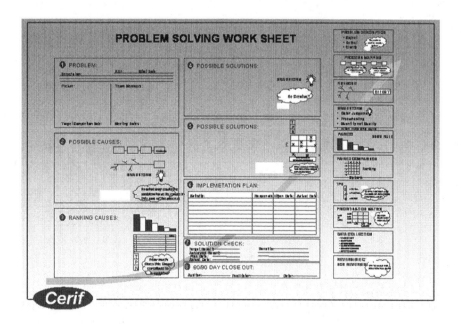

Figure 2.10   Project storyboard

Pause for thought

Does your organisation have any form of organised small group activities If not, what might be the reasons? If you do have them, are they working very well, do people seem to be excited and enthusiastic, are the teams properly supported. If you are sceptical about this concept might you go to see it for yourself? Most Japanese companies have quality circles.

Communication and recognition

In all organisations where success is known to have been achieved, considerable efforts have been made to give recognition to successful teams and to enable all of them to display their work. Typically, story boards are erected at convenient locations near to the work areas to enable the teams to post their charts and other examples of their work. This is partly for recognition purposes but also to encourage others to offer suggestions to the teams.

Displays of completed projects are posted including photographs of the teams and any awards that they may have received. All this activity is to demonstrate commitment and to ensure that the improvement concept has the highest possible profile in order to maintain the highest possible level of consciousness as to the importance attached to these activities.

Statistical process control

The first concept to be introduced to QC circles following the establishment of the initial problem-solving tools was statistical process control. The potential power of this concept had been well illustrated in the late 1950s by Dr Taguchi who received the Deming Prize for Literature for his work. He was able to demonstrate by use of what he defined as being the 'quadratic loss function' that all variability no matter how rare a loss to society and that this variation is, can be continually reduced through the use of SPC.

Statistical process control (in those days it was known as statistical quality control even in the West) was already well known to the Japanese even before World War II. However, the Japanese, as was the case in the West, were not able to introduce it effectively, although they had some success in the telecommunications industry in 1945 partly due to pressure from General MacArthur who was pressurising the Japanese to adopt American management methods. Training the QC circles in these methods proved a breakthrough.

The industrial power unleashed partly by Professor Ishikawa's observations but supported by the work of other activists such as **Homer Sarasohn et al., Dr Deming, Dr Juran** and **Dr Feigenbaum** propelled Japan into world leadership in both quality and productivity.

The West paid very little serious attention to this phenomenon until the late 1970s. There then followed a decade of panic by Western companies, some were driven to extinction, other saw their market shares plummeting whilst at the same time even technological leadership in some sectors, especially consumer electronics and automotive production, steadily shift to the East.

Pause for thought

Do any of the ideas suggested so far make you feel that you need to take some action. Before we move on, please write down everything that you can think of that seems relevant. If you come up with a good plan, maybe I will buy shares in your company!

Summary of learning

In this section you will have learned:

- Definition of quality circles
- History of quality circles
- The essence of quality circles
- Quality circles and the PDCA cycle
- Developments from quality circles
- Advantages of adopting quality circles
- Other types of performance-improvement teams
- Storyboards
- Statistical process control
- Impact of quality circles on the West.

SECTION 7: QUALITY LEADERS

What you will learn in this section:

- Influence of the quality gurus on work organisation
- Organisation support for team activities
- American gurus
- Leading Japanese gurus
- The United States' quality gurus in detail – Deming, Juran and Crosby.

There were many quality and behavioural experts who each had significant influence on the post-World War II re-creation of modern Japan, some at least as important as those mentioned here in many respects, but the following are the most frequently mentioned.

- **Professor Ishikawa** – was responsible for combining craftsmanship with division of labour to create synergy – already covered in the section above.
- **Dr Deming and Phil Crosby articulated 14 points. Deming's were created in the 1980s** to identify the key features developed by the Japanese to create a new system of management and Crosby's also in the 1980s to explain his proprietary approach to quality management.

- **Juran's** managerial breakthrough (covered in Chapter 4) and his concept for 'overcoming resistance to change'.
- **Feigenbaum's** ground breaking concept 'quality responsibility'.

Later in this chapter and subsequent chapters we will discuss:

- Steering committees, quality councils and other considerations for the support of participative approaches
- Facilitation and management support for team-based improvement activities
- Training facilitators, team leaders and team members.

Dr Taguchi

Dr Taguchi was an eminent statistician and he developed many concepts used in quality improvement including a clever shortcut method for design of experiments, but before all of that he was awarded the Deming Prize for Literature with his famous quadratic loss function in the 1950s. Mentioned in the section above, this term might sound a bit daunting but the theory is simple and demonstrates conclusively that all variation is costly. It was this observation that led the Japanese on a relentless quest for improvement by any means in all aspects of industry.

In the early 1980s the West, and especially the USA, woke up to the importance of quality and its core role in the sudden explosion of Japanese economic power. There followed an eruption of interest in the development of post-war Japanese industry. People became interested in the quality gurus and Dr Juran, Dr Feigenbaum (who first coined the term total quality control), Dr Deming and Phillip Crosby were catapulted into the limelight.

Dr Deming hurriedly wrote a book entitled *Out of the Crisis*, in which he enunciated 14 points that were at the basis of Japan's success. Erroneously, many people in the West believe that he taught these points to the Japanese, which is quite wrong. Many of them were the result of Japanese creativity, some were attributable to Professor Ishikawa, some to other Japanese business people and

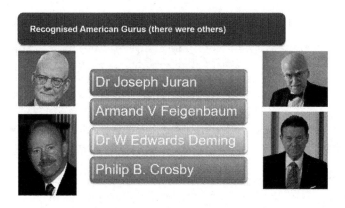

Figure 2.11   American quality gurus

academics and others to Western management. It could be said to be 'East–West fusion'.

1. Create constancy of purpose toward improvement of product and service, with the aim to become competitive and to stay in business, and to provide jobs. (People needed to be told that!)

2. Adopt the new philosophy. We are in a new economic age. Western management must awaken to the challenge, must learn their responsibilities, and take on leadership for change. (OK but what is the new philosophy that he is referring to? Of course, it is the Japanese approach and Deming had very little to do with the creation of that – good advice but hardly earth shattering by the time he articulated it.)

3. Cease dependence on inspection to achieve quality. Eliminate the need for inspection on a mass basis by building quality into the product in the first place. (I'm not sure Deming was even born when this idea became known.)

4. End the practice of awarding business on the basis of price tag. Instead, minimise total cost. Move towards a single supplier for any one item, on a long-term relationship of loyalty and trust. (The Japanese were doing that long before the 14 points were published.)

5. Improve constantly and forever the system of production and service, to improve quality and productivity, and thus constantly decrease costs. (This was supposed to be a new idea!)

6. Institute training on the job (this was at the core of Japanese people development programmes from the start).

7. Institute leadership. The aim of supervision should be to help people and machines and gadgets to do a better job. Supervision of management is in need of an overhaul, as well as supervision of production workers.

8. Drive out fear, so that everyone may work effectively for the company. (This could well have been his observation. It is a good one if it is.)

9. Break down barriers between departments. People in research, design, sales, and production must work as a team, to foresee problems of production and in use that may be encountered with the product or service. (This was known and recognised as a problem long before Deming noted it – it was known as Silo Management.)

10. Eliminate slogans, exhortations and targets for the workforce asking for zero defects and new levels of productivity. Such exhortations only create adversarial relationships, as the bulk of the causes of low quality and low productivity belong to the system and thus lie beyond the power of the work force. (Taken from Juran on worker-controllable errors.)

11. Eliminate work standards (quotas) on the factory floor. Substitute leadership.

12. Eliminate management by objective. Eliminate management by numbers, numerical goals. Substitute leadership.

13. Remove barriers that rob the hourly paid worker of his right to pride in workmanship. The responsibility of supervisors must be changed from sheer numbers to quality. Remove barriers that rob people in management and

engineering of their right to pride in workmanship. This means, inter alia, abolishment of the annual or merit rating and management by objective. (This goes all the way back to Douglas McGregor's famous Theory X/ Theory Y.)

14. Institute a vigorous program of education and self-improvement. (This was happening in Japan long before Deming articulated his 14 points.)

15. Put everybody in the company to work to accomplish the transformation. The transformation is everybody's job (this was originally Feigenbaum's idea).

Dr Juran is, and always was, the most prolific author of quality management books of all time. He had been writing books constantly since even before World War II and it was his book *Managerial Breakthrough*, which was first published in the 1950s, that sparked much of the Japanese quality revolution and is the foundation of Six Sigma and other systematic approaches to business performance improvement.

Dr Juran's breakthrough process

Dr Juran's breakthrough process results in solutions to problems, which must then be implemented. He identified that there are two kinds of possible solution, those that are reversible, in other words, things could slip back to the way that they were, and those that are irreversible where it is impossible for the problem to reappear at least not for the same reasons as previously. An example of a reversible solution is one where perhaps an operator is retrained to perform a task in a way that is different from previous practice. After the retraining, the problem is resolved but perhaps it comes back due to the worker going back to the previous practice.

An example of an irreversible solution might be the redesign of a form to eliminate the need for handwriting. Once the solution has been implemented and the old forms destroyed, it is then impossible for the problem of misreading handwriting to reoccur. However, Dr Juran discovered another problem in the implementation process called 'resistance to change'. It is often noted by those who advocate change that it is not always easy to get the proposals accepted even when the change is beneficial to the workers themselves.

Change is a possible threat to some things that might be important to those affected. The change, therefore, will be resisted until it has been tested as to whether it impacts on these criteria. Until this has been done the proposers will be frustrated and possibly confused but if they continue to push, the resistance digs in and conflict might ensue. The change might be resented due to lack of consultation. There might be valid objections to the proposals, which are known to those affected but maybe not to the proposers.

Overcoming resistance to change

- Involve those affected and give recognition
- Give people time to adjust to the proposals and to raise possible objections
- Work with the leadership of the culture. Convincing the leadership will often make it easier to convince the rest of the group.

Pause for thought

In your workplace, think of some important change that was introduced. How was it introduced to the workforce? Was it accepted? Was it initially rejected even though it was intended to be beneficial? Can you pin point the possible reasons?

Prior to Dr Deming's book *Out of the Crisis*, another American guru, Philip Crosby, had also written a book entitled *Quality is Free*. This also included 14 points, but these were different from Dr Deming's. Crosby was a powerful marketer and was able to get himself listened to.

1. Management is committed to quality – and this is clear to all: Clarify where management stands on quality. It is necessary to consistently produce conforming products and services at the optimum price. The device to accomplish this is the use of defect prevention techniques in the operating departments:

- Engineering
- Manufacturing
- Quality control
- Purchasing
- Sales and others.

2. Create quality improvement teams – with representatives from all workgroups and functions: These teams run the **quality improvement** programme. Since every function of an operation contributes to defect levels, every function must participate in the quality improvement effort. The degree of participation is best determined by the situation that exists. However, everyone can improve.

3. Measure processes to determine current and potential quality issues: Communicate current and potential non-conformance problems in a manner that permits objective evaluation and corrective action. Basic **quality** measurement data is obtained from the inspection and test reports, which are broken down by operating areas of the plant. By comparing the rejection data with the input data, it is possible to know the rejection rates. Since most companies have such systems, it is not necessary to go into them in detail. It should be mentioned that unless this data is reported properly, it is useless. After all, their only purpose is to warn management of serious situations. They should be used to identify specific problems needing corrective action, and the quality department should report them.

4. Calculate the cost of (poor) quality: Define the ingredients of the COQ and explain its use as a management tool.

5. Raise quality awareness of all employees: Provide a method of raising the personal concern felt by all personnel in the company toward the conformance of the product or service and the quality reputation of the company. By the time a company is ready for the quality awareness step, they should have a good idea of the types and expense of the problems being faced. The **quality measurement** and COQ steps will have revealed them.

6. Take actions to correct quality issues: Provide a systematic method of permanently resolving the problems that are identified through previous action steps. Problems that are identified during the acceptance operation or by some other means must be documented and then resolved formally.

7. Monitor progress of quality improvement – establish a zero defects committee: Examine the various activities that must be conducted in preparation for formally launching the **zero defects program** – The **quality improvement** task team should list all the individual action steps that build up to zero defects day to make the most meaningful presentation of the concept and action plan to personnel of the company.

These steps, placed on a schedule and assigned to members of the team for execution, will provide a clean energy flow into an organisation-wide **zero defects** commitment.

Since it is a natural step, it is not difficult, but because of the significance of it, management must make sure it is conducted properly.

8. Train supervisors in quality improvement: Define the type of training supervisors need in order to actively carry out their part of the **quality improvement** program. The supervisor, from the board chairman down, is the key to achieving improvement goals. The supervisor gives the individual employees their attitudes and work standards, whether in engineering, sales, computer programming, or wherever.

Therefore, the supervisor must be given primary consideration when laying out the program. The departmental representatives on the task team will be able to communicate much of the planning and concepts to the supervisors, but individual classes are essential to make sure that they properly understand and can implement the programme.

9. Hold zero defects days: Create an event that will let all employees realise, through personal experience, that there has been a change. Zero defects is a revelation to all involved that they are embarking on a new way of corporate life. Working under this discipline requires personal commitments and understanding. Therefore, it is necessary that all members of the company participate in an experience that will make them aware of this change.

10. Encourage employees to create their own quality improvement goals: Turn pledges and commitments into action by encouraging individuals to establish improvement goals for themselves and their groups. About a week after **zero defects** day, individual supervisors should ask their people what kind of goals they should set for themselves. Try to get two goals from each area. These goals should be specific and measurable.

11. Encourage employee communication with management about obstacles to quality (error-cause removal): Give the individual employee a method of communicating to management the situations that make it difficult for the employee to fulfil the pledge to improve. One of the most difficult problems employees face is their inability to communicate problems to management. Sometimes, they just put up with problems because they do not consider them important enough to bother the supervisor. Sometimes supervisors don't listen anyway. Suggestion programs are some help, but in a suggestion program the worker is required to know the problem and also propose a solution. **Error-cause removal** (ECR) is set up on the basis that the worker need only recognise the problem. When the worker has stated the problem, the proper department in the plant can look into it. Studies of ECR programs show that over 90% of the items submitted are acted upon, and 75% can be handled at the first level of supervision. The number of ECRs that save money is extremely high, since the worker generates savings every time the job is done better or quicker.

12. Recognise participants' effort: Appreciate those who participate. People really don't work for money. They go to work for it, but once the salary has been established, their concern is appreciation. Recognise their contribution publicly and noisily, but don't demean them by applying a price tag to everything.

13. Create quality councils: Bring together the professional quality people for planned communication on a regular basis. It is vital for the professional quality people of an organisation to meet regularly just to share their problems, feelings, and experiences, with each other. Primarily concerned with measurement and reporting, isolated even in the midst of many fellow workers, it is easy for them to become influenced by the urgency of activity in their work areas. Consistency of attitude and purpose is the essential personal characteristic of one who evaluates another's work. This is not only because of the importance of the work itself but because those who submit work unconsciously draw a great deal of their performance standard from the professional evaluator.

14. Do it all over again – quality improvement does not end: Emphasise that the quality improvement program never ends. There is always a great sign of relief when goals are reached. If care is not taken, the entire program will end at that moment. It is necessary to construct a new quality improvement team, and to let them begin again and create their own communications.

Dr Feigenbaum

Dr Armand Feigenbaum postulated (and we grossly paraphrase) that quality should be everybody's responsibility. For example, quality control does not hold the pencil of the designer, turn the handles on the machines, answer customer calls in the office and if these people do not care, if they just do their jobs according to instructions then no amount of quality control will achieve the required results.

He also said that if quality is everybody's job then it becomes nobody's job.

Therefore, it is the job of quality assurance and quality control not to be responsible for quality but to support the work of those who are. If you obtain a copy of his book *Total Quality Control* you will find that it contains a grid of quality responsibilities related to each of the key functions of the business.

Pause for thought

From the work of the gurus could you identify, say, four concepts that each of them highlighted that you think might be relevant to your organisation?

Summary of learning

In this section you will have learned:

- Think of the influence of the quality gurus on work organisation
- Organisational support for team activities
- About American gurus
- About leading Japanese gurus
- The West's quality gurus in detail – Deming, Juran and Crosby

SECTION 8: THE STEERING COMMITTEE

What you will learn in this section:

1. Why a steering committee?
2. Forming a steering committee.
3. The work of a steering committee.
4. Top management involvement.
5. Middle management and specialists.
6. Supervision.
7. General responsibilities of steering committees.

Why steering committees?

Consider this statement by Professor Ishikawa describing Japanese total quality control (TQC).

In Quality Control work, all *persons concerned from the President down to the operators have received suitable education and training. In one word, we have practice, so to speak, quality work of, by and for all.*

(*What is Total Quality Control?* 1985)

This led him to describe such a structure as 'company wide quality control'. To achieve it, organisations have found it beneficial to establish what are often called 'steering committees'.

If because of this book you are able to implement QC circles, we hope that the following will be a useful source of information to support the development of a successful programme.

It would be impossible to commence the successful implementation of fully participative management, before all people likely to be affected have been given an opportunity to decide for themselves whether they want to become involved. To ensure a positive response at all levels, it would be necessary to give awareness presentations to top management, middle management, trade union representatives, specialists and supervisors.

If the general reaction is favourable, and there are no major objections, it is then possible to draw up an action plan for the development of a companywide participative programme.

Getting started: typically, the chief executive, a director or high-level manager will have to make the decision that this is desirable, probably following a presentation by an expert talking to top management.

After this presentation, one or two other directors or senior managers may have taken an active interest, and so a small team will have been formed. They will plan the next steps.

The steering committee is not a substitute for the normal top management meetings. These would take place as usual. However, experience shows that when there is a move from the traditional Western management approach to participative quality leadership, the number of new issues involved would greatly overload the agenda for standard management meetings. This being the case, many items would fall off the agenda due to time constraints. Because what we are proposing here is a major change programme, it makes sense to have its own agenda and a revised list of attendees to represent the various interests.

Membership of the steering committee is an official appointment as is any other appointment for the running of the organisation. Not all the participants will necessarily have put themselves forward, they may have been invited but can also volunteer.

The steering committee can, if properly constructed, very effectively create the necessary consensus style characteristics essential for the development of a healthy participative programme and these can be linked to the down flow of ideas from the *Hoshin Kanri* process. *Hoshin Kanri* is the name given to a totally integrated management system in which the power of the whole is greater than the sum of the contributing parts as in the case of the triangles shown in Figure 2.1. *Hoshin Kanri* is only given light treatment in this book but is covered fully in the complementary book *Hoshin Kanri a Strategic Approach to Continuous Improvement*.

Diagonal slice

If the steering committee is reasonably representative of all interests, it will be 'in touch' or 'wired in' to the feelings of all members of staff. It will also induce confidence. For example, shop floor workers in a factory who are worried about some aspect of QC circles may find it difficult to talk to a senior manager or someone from another section or department. However, if there is someone on the steering committee at their own level, or who they find accessible, then they will have no worries about approaching that person with their observations. Not only is this important from that individual's point of view, it is also valuable to the steering committee itself, because now, they will be more sensitive to the general mood and to people's perspective of the QC Circle programme.

Top management team

A steering committee will be largely ineffectual if it does not contain as a member one of the ultimate decision makers at site level. Someone who can sign cheques is a reasonable guide usually. Otherwise, the steering committee will lack authority, and all the important decisions will be made by a third party, i.e. top management. It also means that the committee will be required to make representations to top management for decisions to be made and top managers may not appreciate the importance of requests made, simply because they have been less involved. It is a tremendous boost to everybody's confidence to see the active membership of the steering committee be the chief executive. This is the most impressive manifestation of 'management commitment'.

Middle management and specialists

This level or group is usually the linchpin of a participative programme. The attitudes of middle managers both individually and collectively will determine to a large extent the 'flavour' of that company's QC circle activities. If care is not taken to develop middle managers at the same time as the newly formed circles, some managers may become afraid of the growing confidence of their people. Also, through the acquisition of skills by the circles, some managers may become nervous if their people appear more competent than they themselves. Key middle managers useful to the development of the programme therefore, are those who choose to become more closely associated with the QC circles and can therefore play a vital role in the steering committee activities.

Supervision

In some companies the role of the supervisor, group leader or foreman is more clearly defined than in others. It is just as important that someone from this level should find a place on the steering committee. Some supervisors may be nervous about how circles will affect their role, and it would be dangerous if the steering committee was not sensitive to these feelings and neglected to have them represented.

Appreciation

The steering committee should always ensure that all circles receive adequate recognition for their achievements or some circles will feel more appreciated than others. The steering committee should be aware of this and can help overcome any difficulties by giving some circles more exposure in news sheets, etc. The aim should always be to keep the whole programme at the same healthy level.

Liaison

The circle programme does not exist in isolation from other company activities, some of which may overlap the work of circles. In the context of organisational development, it may be necessary for the quality circle steering committee to interface with other groups concerned with other concepts such as task force and project group activities in order to produce an integrated programme.

Development

As the programme develops, the ultimate power of quality circles may be realised when circles become an integral part of company activities – in other words, when everybody can be a member of a circle. The ultimate aim should be 100% membership, and quality circles are simply the way a company manages its people.

The people in charge of such functions as quality control and quality assurance should realise the importance of quality circles in helping the achievement of their own objectives. To do so they will be feeding information to the groups. This information will include customer complaints data, articles from quality journals and anything else that quality control thinks will enhance circle activities.

This also applies to other functions such as production engineering, work study, accounts, etc. The steering committee can play an extremely important role in the encouragement of such developments.

Summary

Those organisations that have already begun circle programmes but have not yet established steering committees would do well to start such committees immediately. One well-known American company did not do this, and the programme collapsed when the facilitator left. Fortunately, it has since been re-established, but they now have an active steering committee.

Facilitation

The facilitator should always be a member of the steering committee and will usually be the main source of information. In a larger company with several facilitators, it would be unreasonable for all of them to be represented. Usually, one of them, who in this case could be termed a coordinator, will probably be more senior than the others. The coordinator would be their steering committee representative and would report back to the other facilitators at a facilitators' meeting.

The steering committee will be the facilitators' main source of support. In companies that have a participative, consensus style of management, this support will not be so important, but in others, particularly those which suffer from strong interdepartmental rivalries, the support of the steering committee may be the difference between success and failure. This is further evidence of the value of having a broadly based steering committee

---

Pause for thought

If you are a member of or have good access to the top management team in your organisation and by the time you have decided to implement many of the ideas in this book and it appears to be a lot of work, we seriously advise you to consider setting up a steering team to oversee the changes and I would venture to suggest that all proposed members have the opportunity to read this book first!

Summary of learning

What you have learned in this section:

- • Why a steering committee?
- • Forming a steering committee
- • The work of a steering committee
- • Top management involvement
- • Middle management and specialists
- • Supervision
- • General responsibilities of steering committees.

SECTION 9: TEAM LEADERSHIP

What you will learn in this section:

- • About this section
- • Choice of team leader
- • Responsibilities of a team leader.

This is a short section with only two topics, but experience suggests that these two items are of great importance to the success of the programme.

We have tended to concentrate on the choice of leader for groups such as quality circles mainly because membership of such teams is voluntary not mandatory. Therefore, the choice of leader of such teams is critically important. People will vote with their feet if they do not like the leader!

We have discussed the role of various types of team and leadership in general. However, because self-selecting teams need leadership and also because the membership of such teams is voluntary, who leads the group is an issue that needs careful thought.

When teams choose their own leader, they usually select the person who they think is most likely to take them where they want to go. It is usually, but not always, the most popular person or perhaps the one they have most confidence in or the one who knows the most. The choices vary according to the situation.

For example, if the group were stuck high up on a mountain in a blizzard, popularity would be a minor consideration. The group would almost certainly select the one whose mountaineering knowledge was most likely to get them to safety regardless of his popularity. If it were for the organisation of a social function, the choice of leader would probably be different again.

In the case of imposed leaders, the criteria are different. In this case the person chosen as leader will be the one thought to be most likely to achieve the goals of those who chose him and that might not necessarily be what the group themselves want. Since quality circles are voluntary, the choice might affect the enthusiasm to participate. Industrially, this is the more common situation.

Choice of team leader

Of course, if the choice is really in conflict with the aspirations of the group, then some or all the participants might decide to drop out. Therefore, the selection of a good leader is paramount to the success of the project and to gain the maximum support from the group. Good leaders exhibit certain qualities that are common sense if considered but this is not always the case! An essential part of getting results through people is good leadership.

Responsibilities of the team leader

Work with the group to:

- Gain consensus in the identification of the aims and vision for the department, and its purpose, and direction – and define the activities involved (the tasks)
- Do not forget, that what might appear to be an attractive idea to one person might be seen completely differently by another so consensus is important
- Identify resources, people, processes, systems and tools (including financials, communications, IT)
- Create the plan to achieve the tasks – deliverables, measures, timescales, strategy and tactics
- Establish responsibilities, objectives, accountabilities and measures, by agreement and delegation
- Set standards, quality, time and reporting parameters
- Support and maintain activities against agreed parameters
- Monitor and help to ensure maintenance of the overall performance of the group against their agreed plan
- Report on progress towards the group's aim
- Review re-assess, adjust plan, methods and targets as necessary

- Establish, agree and communicate standards of performance and behaviour
- Establish style, culture, approach of the group – soft skill elements
- Monitor and maintain discipline, ethics, integrity and focus on objectives
- Anticipate and resolve group conflict, struggles or disagreements
- Assess and change as necessary the balance and composition of the group
- Develop team-working, cooperation, morale and team-spirit
- Develop the collective maturity and capability of the group – progressively increase group freedom and authority
- Encourage the team towards objectives and aims – motivate the group and provide a collective sense of purpose
- Identify, develop and agree team – and project-leadership roles within group
- Enable, facilitate and ensure effective internal and external group communications
- Identify and meet group training needs
- Give feedback to the group on overall progress; consult with, and seek feedback and input from the group understand the team members as individuals – personality, skills, strengths, needs, aims and fears
- Assist and support individuals – plans, problems, challenges, highs and lows
- Identify and agree appropriate individual responsibilities and objectives
- Give recognition and praise to individuals – acknowledge effort and good work

- Where appropriate, reward individuals with extra responsibility, advancement and status
- Identify, develop and utilise everyone's capabilities and strengths
- Train and develop individual team members
- Develop individual freedom and authority
- Support facilitators of team working.

Generate enthusiasm

- Remember – membership is voluntary in self-selecting teams
- Be open about the activities of the team
- Involve all the members
- Don't be too ambitious
- Avoid pay and personalities
- Use the problem-solving techniques
- Look towards solutions.

Leading teamwork for success

We have mentioned earlier that every so often we see a team of not particularly good sportsmen winning a major trophy and beating rivals whose individual skills and resources are much greater than the team in question.

What is behind all of this? Well, when we dig in, we find that whilst these events appear to happen at random there are, in fact, some factors that come up repeatedly in such situations. In this section we will explore these important factors.

Figure 2.12 shows the typical performance of teams from poor performing to top class.

This statistic generally happens by pure luck. Of the typical teams, they range from near failure to reasonably good but with an average of 'not impressive'. This result is typical of appointed committees

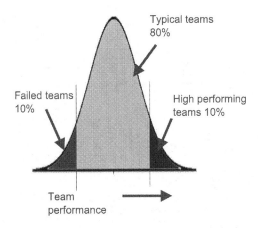

Figure 2.12   Team performance

What we want to do is to make happen on purpose what has generally happened largely by luck – to be a highly performing team. The following figures will give you some good clues on how to do this but remember, there is never an end to finding ways to be even better

So, what we can learn from this is that what we are saying does not apply just to work. The ideas can be applied in any situation where a group of people set out to achieve some sort of goal, for example, raising the funds for a new roof for the village hall, organising an event, being a member of a sports team, etc. Some of the ideas can apply even to individual efforts.

Some time ago, a researcher named Dr Bruce Tuckman studied this subject and came to the conclusion that there was a mechanism by which we can recognise what is going on during teamwork and manage it in order to produce high-performing teams.

Dr Tuckman referred to it as forming, storming, norming and performing. Dr Tuckman's theory is relevant to all types of group, including committees, but that is not the end of the story. It is also the case that some teams can consistently produce better results than others even when all are using Tuckman's approach.

Pause for thought

Before reading on through the rest of this section, think back to some past experiences you may have had on a successful team. It could be one at work or in your home life – something that might have been a once in a lifetime experience. You won a sports challenge, raised a lot of money – anything.

Dr Meredith J. Belbin's theories

One possible reason for the even greater success of some Tuckman-based teams over others is some ideas that were unearthed by Dr Meredith J. Belbin, a social scientist.

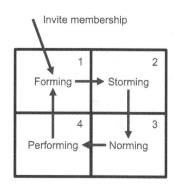

Figure 2.13   Tuckman's theory

Figure 2.14   Dr Meredith J. Belbin

He discovered that there were some personality characteristics that we all share in varying proportions, and that, if identified, teams can be constructed that will almost always win.

Dr Belbin was a management scientist who originally identified eight key characteristics (later increased to nine) that people can exhibit in team situations. These vary from individual to individual and whilst some people might be strong in one or two of the characteristics, they will be weak in others. A team that has good balance of each of these is likely to succeed over teams that are less well balanced.

We think that when you have looked at the characteristics of each you will easily recognise both yourself and others around you as to which are predominant in their nature. There is no right or wrong, good or bad because as a group we need all of them to be successful.

- Implementer (IM)
- Coordinator (CO)
- Shaper (SH)
- Plant (PL)
- Resource Investigator (RI)
- Monitor Evaluator (ME)
- Team Worker (TW)
- Completer-Finisher (CF)
- Specialist (S).

More details on these and other variants can be found in Dr Belbin's work, which is ongoing and dynamic. Undoubtedly, it will continue to develop during the lifetime of this book.

For copyright reasons we have only provided an overview of the Belbin approach here. To find out which team roles you naturally fulfil, or to profile your team, visit www.belbin.com. The simplest approach is to use Belbin's 'Self Perception Inventory'. It is fun to use, and very interesting to see your own and your colleagues' characteristics.

Whilst there are other approaches, the attraction of the Belbin approach is that it is completely harmless. There are no 'bad', character types exposed so there is nothing to be afraid of and it really does help in high performing team development

Other behavioural factors

Quite apart from all this behavioural theory, there are many other factors that usually characterise high-performing teams which we will briefly mention here.

Really talk and really listen

Members of high-performing teams are noted for the fact that they listen intently to what any other team member is saying.

Watch body language

Members not only listen but they watch as well. Body language is vitally important. The team can only perform at the highest level if every member is on top form. Maybe one person is upset about something. Watch for the signs.

Care and maintenance

If a member is unhappy, it is important to try to get to the bottom of it and, if possible, get them back to being happy, then they will again perform well. This situation can occur at any time and with any team member so be on the watch out.

Celebrate success

There are an almost unlimited number of ways that teams can do this. It might involve going to the pub, but it doesn't have to be. Sometimes, the banter that takes place after an event can be in the changing room, analysing the game if it was a game.

Use humour constructively

Successful teams use humour a lot. Leg pulling is very common but be careful, leg pulling can also be destructive too if it carries poisonous messages to convey some form of criticism. Successful teams do not do this. We are back now to **care and maintenance**

If there is an issue regarding a team member remember Tuckman's 'storming' idea. It is OK to storm if it is followed by 'norming'. The storm gets everything out in the open. Everyone can learn from the situation **if** they 'really talk and really listen'.

More on humour

Another feature of successful teams is 'in jokes'. This can be quite off putting for someone outside of the team, but it has a powerful bonding effect on team members. Sometimes, the jokes are complete nonsense to anyone who was not part of the group.

New members

Remember, if a new person joins a group, they need to be brought up to speed on all of the 'in jokes', war stories etc. to make them feel they are accepted but the same goes the other way. Someone leaving a group as well as someone joining is disruptive to the sense of camaraderie. Again, the Tuckman concept is useful. Also, the team must be mindful of the effect on the Belbin balance and make adjustments accordingly

---

Pause for thought

How good is your organisation regarding team dynamics? If you do not use it, you might consider checking out how teams might be established based on Belbin and Tuckman's theories.

The good thing about both approaches is that nobody gets hurt and people love to know their Belbin characters and there are no bad ones. There are psychosomatic tests that can be dangerous when in the wrong hands. Maybe some are just dangerous anyway, so caution is a good idea in this field.

Conclusion: what comes next?

If these basic rules are followed, not only will teamwork achieve outstanding results, but it will be huge fun. But a word of caution. Members must keep all this balance with the rest of their lives, their family, friends, work colleagues and relatives. It is easy to get so carried away with the team project that other important relationships get overlooked.

Also, most team activities come to an end sometime or, if it is a sport, it might be seasonal. There can be quite an anti-climax after a heady experience and team members might have problems in readjusting their lives. It is suggested that the team see this as part of the process and it is not over until everyone feels grounded. One way to do this is to involve loved ones is some sort of celebratory event. This can help with the grounding process and make our friends feel involved.

SECTION 10: COMMUNICATIONS

In this section you will learn about:

- What is communication
- Verbal communication
- 1 to 1 vs 1 to many
- One way
- Non-verbal
- Lack of communication
- Listening skills.

These days when the word 'communication' is mentioned we tend to think immediately of satellites, electronics, high-speed data streaming, etc. but these are only 'means' of communication not communication itself. In this section we will be dealing with the subject in its person-to-person form – what it means to both you and me.

Communication is the process of passing information from one person or group to another. Sometimes, this is a one-way process where no response is intended. In other cases, it is a two-way process as in conversation where part of the purpose of the communication is to obtain a response. We all complain about poor communications and we express this concern in many ways.

For example, we are all familiar with the following:

- Nobody tells me anything
- If you had a problem, why did you not say so?
- If you wanted to know you could always ask
- My door is always open if there is anything you want to say.

Or, domestically:

- Oh no. I asked you to switch the cooker on at 11.00 why did you not listen?
- I thought I told you to put the cat out before going to work?

Always remember: **really talk and really listen!** Try to be interesting and try to absorb what you are being told.

Verbal is not only vocal. It includes all tangible forms of communication such as the written word (this text is an attempt by us to communicate with you). This is also 'one-way' communication but as such is not so much communication as 'transmission' because at this point it is only one way.

Figure 2.15   One-to-one communication

Apart from the spoken word, and even then, in some situations, unless there is the opportunity for feedback and then, a further response based on that feedback, communication is not really taking place. One-way communication yes, but that is not true communication.

There is also one-to-one communication, this is usually speech either face-to-face or by telephone but today it could arguably include text messages and other electronic forms. Then there is one to many. This might be addressing a rally, making a presentation, etc. Each has their uses according to the situation but making the decision of which to use will depend upon the desired outcome.

As has been said, one-way communication is not completely satisfactory on its own because without feedback it is impossible to know whether communication has taken place at all. This is true for television and radio broadcasts, photographs, video, books, emails (before a response from the addressee), text messages, faxes, advertisements, notice boards, etc. All of these are one way where the sender has no idea as to the reaction of those who are on the receiving end. Not until, that is, the transmitter carries out a survey or solicits and receives a response.

Think of some one-way communications that take place in your work. Write them down if necessary. In some cases, they may be satisfactory, for example a notice saying that a particular toilet will be closed for cleaning between say 10.00 and 10.30 a.m. on Tuesday. A notice by a motorway may say it will be closed for repair from 10.00 Wednesday 5 to 10 June. Apart from the fact that frequently the text on these notices is too small to read when travelling at speed, the information is probably inadequate. You will want to give some feedback, possibly to be fully informed or sometimes to change or influence the decision.

In terms of communication within an organisation, all these forms of communication have their uses but all of them have their relative merits and possibly pitfalls. In your organisation look at the various forms of communication that impact on you and ask yourself if you think they are satisfactory and how they might be improved.

This usually refers to 'body language'. Generally, when we are in the company of others, we communicate whether we choose to or not. Much of this is unconscious where the communicator is often unaware that they are communicating!

Non-verbal communication

For example, boredom is shown in many ways. Appearing to not be paying attention is one of the most obvious and it often deliberate. For example, picking at one's finger nails, making a point of looking at the watch, staring up at the ceiling, sitting with tightly folded arms and staring straight ahead is both a sign of boredom and a hostile demonstration of not wanting to be there.

Body language can be both subtle and very obvious. The observer must be aware of these tell-tale signs and react accordingly. In teamwork, recognising body language and reacting to it quickly can make the difference between success and failure.

Probably the two most frequently stated possible causes of problems in brainstorming sessions are 'lack of training' and 'poor communications'. From the work that you have done so far in this chapter we feel sure that you will have few doubts as the likely reasons for that. Broken links in the chain of PDCA loops is the most

Figure 2.16   Body language

common. People at one level do not always appreciate the importance of communicating with people at another level for several reasons:

Humans tend to be tribal and therefore do not feel comfortable with people they perceive to be above or below them or even from different departments. Also, they do not always know how to deal with them. For example, upper management usually speaks what might be described as the 'language of money'. Most of their concerns are couched in terms of 'return on investment', 'profit', 'asset ratios', etc.

One of the biggest problems in organisations is the lack of real communication between departments. Designers talk to designers, operations talks to operations, etc. Not only do they not talk freely across these boundaries, they do not even use the same terminology so even when talking is attempted the message is not always correctly understood. So, what happens is that the designer designs something and then throws it over the wall into operations who then must do what they can to interpret the requirement. Most of the time they do but in too many cases they do not, with the inevitable consequence of quality-related costs.

Frequently, the message gets lost because the other person does not speak the same work language. Supervision generally speaks the language of things 'units of scrap', 'number of defects' 'customer complaints' 'absenteeism', and so forth. They and the workforce will express their concerns in the dialect of their respective departments.

Middle management therefore needs to be bilingual. They need to be able to speak the language of money when talking to upper management and the language of things when talking to supervision and the workforce. Most middle managers are skilled in the language of things but not the language of upper management!

As a consequence, vertical communication is generally poor. However, by good use of the PDCA cycle approach, communications can be very much improved.

How often are people heard to say. 'I like him or her because they are good listeners'? Maybe you tell someone something that is possibly important to you but maybe not to them and yet they meet you again some time later and it is clear that not only have they remembered it, but it seems that it was important to them?

A lot of friendships grow out of such experiences. This is just as important in work both between your boss and yourself as it is between yourself and colleagues as well as with those who report to you.

Are you guilty? Well, how often is someone introduced to you and only minutes afterwards we have forgotten their name! Do not worry, we all do it, but it is an indicator that we are not always brilliant listeners.

A good technique that you can practise with your work colleagues is when you are all together in a group, for someone to explain something and then soon afterwards for someone else to be asked to repeat what was said. It does not by any means have to be word for word, just the important points.

Do this a few times and you will be amazed how quickly listening improves. Try it at home with your partner or kids. Some of the worst communication is often between spouses or anywhere within the family.

Pause for thought

A look in the mirror time for most of us! How good are we? What might we do better? Is the organisation fireproof with its communications processes? Where is the notebook?

Summary of learning

In this section you will have learned:

- What is communication?
- Verbal communication
- 1 to 1 vs 1 to many
- One way
- Non-verbal
- Lack of communication

Listening skills.

---

3   The role of monitoring and measuring in making decisions

AIM OF THIS CHAPTER

Most of us spend a lot of our lives monitoring and measuring things but, interestingly, we also make many of our decisions based on opinion and do not bother with data. Sometimes, the data is flawed, anyway. This chapter will explain the importance of data collection and analysis as a means for decision-making with regard to product and service quality, the use and limitations of the broad range of monitoring, measurement and diagnostic tools and techniques available and their application.

The learning outcomes of this chapter are:

- To understand the role of monitoring and measuring in making decisions relating to quality – be aware of the monitoring and measuring techniques that a business might use and where they might be applied
- To understand the use of methods for data collection and analysis – list and describe a selection of analytical tools for diagnosis and control of date. Interpret data and make conclusions or recommendations on results.

FURTHER LEARNING POINTS

All of us have opinions on most things. Frequently, they conflict with the opinions of others. Facts, however, cannot be disputed – if they really are facts!

This chapter shows you how.

We all have opinions of the causes but still the problems remain or if they disappear for a while they re-emerge again later. The reason for all of this is the fact that most problems simply cannot be solved by debate. The only way they can be resolved is to identify theories of causes, this does involve debate, but then we must test those theories to find which are valid and which are not.

We now run up against a series of obstacles. Sometimes, it is not easy to collect data, or it requires what might appear to be complex methods, which may be expensive. It depends how important it is to know. It might be embarrassing in some cases to collect the data. We need results fast, so we continue to rely on our theories. Sadly, we see what we want to see and ignore everything else. The brain works like a filter.

Opinion data tends to highlight the events that suit our preconceived theories and conveniently overlooks data to the contrary. The scary thing is that most of the time we do not even realise that we are doing it. How then, can we be sure that when we collect data it is factual and not misleading? Fortunately, it is not as difficult as it might sometimes seem. Some simple rules can help.

SECTION 1: A FACTUAL APPROACH TO DECISION-MAKING

What you will learn in this section:

- Fact vs opinion
- Seeing is believing, or is it?

Most of us spend a lot of our lives monitoring and measuring things (note the car dashboard in Figure 3.1), so why do we need a chapter dedicated to it? Well, it is true that we do look at data a lot but interestingly we also make many of our decisions based on opinion where we do not bother with data. We all do it. This is OK in many cases and it saves a lot of time, but there are also many cases where even though the decision is important, we still rely on opinion data even though the consequences of getting it wrong could even be catastrophic.

Note that in the picture of the dashboard, there are instruments for both monitoring and for measuring. We use the speedometer mostly for monitoring and except for the use of tachometers in trucks, **we do not always record the data**. It is used for 'in process' control and keeping to speed limits, etc. In the case of the fuel gauge, we might well **use the data for recording**, especially if we are testing the fuel consumption over a period.

Figure 3.1   Dial type instuments

In many cases, such as the car dashboard, we do not need to be very precise when reading the data and there is not much of it anyway so reading it is simple. In contrast, in the cockpit of an aircraft, for example, we need to be very precise when reading and interpreting the data, especially altitude when landing, and with the vast array of dials of all shapes and sizes, reading them is a complex situation. Errors cannot be tolerated!

If it has been necessary to record data such as maybe that for recording the mileage travelled for each tankful of fuel used, we could produce a table of the data as shown here. You would have to study the data very carefully to see if it was telling you anything. Now look at Figure 3.3. It is created from the same the same data.

Miles per tankful.

320, 332, 290, 301, 333, 319, 342, 288,
344, 316, 351, 340, 315, 319, 299, 341,
354, 300, 290, 296, 355, 349, 360, 316,
322, 338, 316, 350, 318, 362,
351, 340, 338, 329, 344, 354, 322, 347,
360, 352, 328, 342

Figure 3.2   Fuel consumption

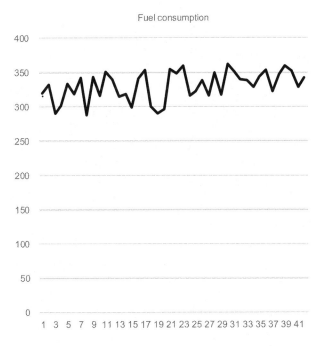

Figure 3.3   Charted data

It was not obvious from the table of data in Figure 3.3 but when put in chart form it clearly shows that there has been a steady improvement in fuel consumption over the period even though the day-to-day results jump up and down.

The reasons could include:

- More careful driving therefore maybe of collecting the data, or
- Possibly generally longer journeys, which are more economical. If a log of journeys had been kept then this theory could be either verified or rejected.

Now there are many more types of data than appear in aircraft cockpits and cars and there are several ways that these data can be managed. In this chapter we look at a range of these and try not only to simplify everything but also hopefully to explain some methods that could well be useful in your work or at home and to help avoid some of the errors that are all too easily made.

Opinion data tends to highlight the events that suit our preconceived theories and conveniently overlooks data to the contrary. We think that you will probably agree that the data on fuel consumption when presented graphically is irrefutable but looking at the gauge swinging about whilst we are driving or in tabular form tells us very little. We can argue as much as we like, but unless we have the data presented in graphic form we will not actually 'know' anything.

Pause for thought

Think about some of the problems in your work that are often debated but never seem to get solved. Make a note of some of the important ones.

Seeing is believing or is it?

Sometimes, it is not easy to collect data, or it requires what might appear to be complex methods, which may be expensive. It depends how important it is to know. It might be embarrassing in some cases to collect the data.

We need results fast, so we continue to rely on our theories. Sadly, we see what we want to see and ignore everything else. This is one of the main reasons why we often have very strong but opposing views to our friends. Remember, however careful we try to be, we can be easily fooled as the following figures will show!

Sadly, we see what we want to see and ignore everything else. The brain works like a filter. There are two faces in Figure 3.4. Which did you see first! Most see the young lady and unless they are given the chance to study it, they will be convinced that this is the only view. How long was it before you saw the old witch?

The next few figures give you some more examples of things not being what they seem. The lesson is that we must be careful with data of all sorts if we want to make good decisions.

We can assure you that in Figure 3.5 the horizontal lines are all straight and perfectly horizontal, but it is hard to believe. This is what we call an optical illusion because the data has fooled the brain.

Figure 3.4   Optical illusions

Are the horizontal lines parallel or do they slope?

Figure 3.5   Fooling the eye and brain

Deliberately misleading information

We need also to be on guard against fake news and the manipulation of data to create deliberately wrong impressions. For example, it might be said that if you take some medicine for a cold it will clear up in seven days. If you do not, then it will drag on for a week.

There is nothing wrong with the data. Here is another we probably all know.

A glass is said to be either half full or half empty depending upon the impression you are trying to give. Always think about what you are being told and do not be fooled.

Mental images

Be careful, as not everybody sees things the same way. For example, a simple word like 'camping' can evoke some very different images depending upon experience.

Summary of learning

We learned that there is very often a large gap between what we believe and reality. Even faced with the facts, there are times where we continue to believe our opinions

We also saw a few optical illusions that take some believing. These were contrived but there are a lot of cases in our work where we are fooled into believing something with our own eyes that convinces us, but it is wrong.

Pause for thought

Is there any misinformation or misleading information in your organisation that comes to mind?

SECTION 2: LIES DAMNED LIES AND STATISTICS

Do not panic! The word **statistics** strikes fear in many people's hearts! But you are much better at statistics than you think. In fact, you probably do not realise it, but you are naturally a statistician.

Example

You would not cross a busy road blindfolded for a bet no matter how big the wager! Why? Because your knowledge of the laws of statistics accumulated through life has taught you that the chance of getting away with it is very low. So, you wouldn't risk it. So, you do understand risk!

You watch a conjuror select a card at random from a pack. It happens to be the ace of spades. So far, you are not impressed. He asks you to shuffle the cards.

He again pulls out the ace of spades and you look surprised. He asks, 'Why the surprise, you were not surprised the first time?' 'No' you might say, 'it had to be one of the 52 so it happened to be that one'.

'Yes, but to do it twice?' You made no calculation but what you have learned in life is that a rare event will happen purely by chance every so often but if the same rare event happens twice, it is less likely to be chance and more likely that the cards were manipulated. I showed the card trick to my cat. He was not impressed even though I pulled the ace out several times!

Our previous cat had got run over. He was not good at statistics either! He had run across the busy road before and without looking. He thought that because he got away with it once he could always do that. Unfortunately, the laws of chance prevailed and exit cat from planet!

If human beings were not naturally and instinctively familiar with the basic rules of statistics, there would probably be no lottery, football pools, casinos or village hall raffles. We all have an instinctive understanding of the laws of probability, at least most of us do. Those who do not, usually end up out of pocket at the local bookmakers!

Collecting most data is simple because, in general, there are only two types, and each can be analysed using some very simple tools. Statisticians will try to blind you with science and make it look difficult but do not be fooled. It is a fact that over 80% of all industrial problems can be solved with the simplest methods.

For the remainder, you can always consult an expert from your local college! However, be wary there, too, as some of them will try to blind you with science in an attempt to practise some of the more sophisticated tools they rarely have the chance to use!

Data comes in a variety of forms and depending upon which will determine the range of possible analytical tools available, as well as how it might be collected.

You are not required to delve that deep into the tools in this book but obviously the more you understand the better!

Much of the time we can use statistical methods to attempt to understand and control variation in processes. In this instance, we are mainly concerned with just two issues, to be able to recognise the difference between:

1. Chance causes of variation each of which are apparently random in nature and generally rare.
2. Those causes of variation that can be identified, predicted and hopefully eliminated. These are called special causes

---

Pause for thought

Is there a culture of fear of data collection and analysis techniques in your organisation? It is very common. People are often mistrustful of the concepts because they do not understand them. Later in this chapter it will be demonstrated that this does not have to be the case and big breakthroughs are possible through the right approach. Make a note to watch out for this.

Summary of learning

Collecting most data is simple because in general there are only two types of data and each can be analysed using some very simple tools.

Much of the time we can use statistical methods to attempt to understand and control variation in processes. In this unit we have been concerned with just two issues: to be able to recognise the difference between chance causes of variation each of which are apparently random in nature and generally fairly rare and those causes of variation that can be identified, often predicted and hopefully eliminated.

SECTION 3: COLLECTING DATA

In this section you will learn:

- Types of data – attribute and variable.
- How much data is required?
- The risks you take with data gathering and analysis.
- Subjective sampling.

Looking at Figure 3.6, you could be forgiven for thinking that data collection is a difficult subject, but we have news for you, none of this is difficult. The reason that it seems difficult is the vast array of different sorts of data we see every day.

It would be easy to imagine that each had its own unique way of dealing with it because it looks different.

We have good news for you. Despite the way it looks, there are in fact only two different sorts of data:

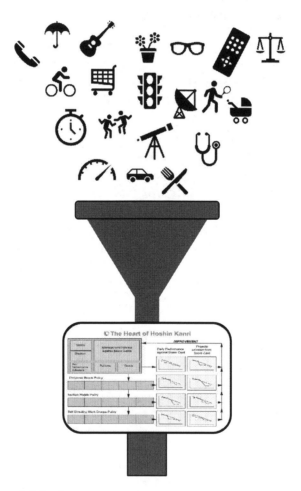

Figure 3.6   Multiple data

- **Attribute** (with the subset subjective)
- **Variable.**

And regardless of which, all the same techniques apply. Later, there is a section on each of these, but fortunately the collection process is very similar and the basic rules much the same.

Attribute data. Basically 'attribute' data means that it always relates to events such as:

- Is/is not
- Something is either good or bad, right/wrong
- Present/absent
- It happened
- It did not happen.

Subjective data (sub group of Attribute data) is so called because it is subjective to the five human senses, smell, taste, sound, feel and appearance, but it is still 'attribute' data. For example: a tea taster tests some tea. Either it tastes OK or it does not, just as with other forms of attribute data.

'Variable' data is what it says it is, variable. For example: distance, speed, voltage, current resistance, weight, etc. Although the data is often analogue in nature, for convenience it is generally digitised for ease of calculation.

How much data do we need?

The answer to this question must always be considered before collecting any data and the answer depends on how accurate we need to be with the results. If we only need to be within a few percentage points (usually the case) then 50 or so samples need to be taken, provided that they are taken at random. However, if higher levels of accuracy are required, it is best to obtain the opinion of a trained statistician. You could check with your local college for some advice

Research has shown that nothing more than the bare basics tools for data gathering and analysis are necessary to deal with more than 80% (some say 90%) of all situations so be wary of anyone trying to sell you complex packages when they might not be needed!

Of course, sampling involves some risk. In fact, there are two special risks that are always present.

1. Saying that something is good when it isn't.
2. Saying something is bad when it is good.

These can be reduced with larger sampling sizes and by always bearing in mind the following additional risks.

Risk 1: bias

The data is biased and gives a wrong impression. This sometimes happens with opinion polls when they are unable to take the opinions of a representative sample of people. This can be caused by both of the following two risks, but the most common cause is not taking the sample at random. Ensuring randomness is one of the most important rules in data collection.

Risk 2: non-repeatability of results

If your results cannot be repeated, it is unlikely that others will necessarily believe the results and the probability is that your theory regarding the probable cause of the problem might be wrong. This is often the case where the test must be destructive. In this case we want the maximum information from the least possible number of tests. Always try to keep the evidence or some non-tested samples from the same population if it is possible.

Risk 3: unrepresentative sample size

Make sure that the sample is big enough to validate the claims being made. For most everyday purposes a sample of 50 will be good enough provided that the data was selected at random. If the results require more precision than say +/– 3 or 4% then this is the time to call on the services of a qualified statistician to help you design the sampling plan. **With subjective sampling, make sure that everyone involved uses the same criteria!**

Suppose we set out to discover how many men in the organisation had beards. Before we rush off to collect the data, let us make sure that everyone involved has the same idea as to how many whiskers actually make a beard. If we do not do that, we might get some strange results. To be sure, we should make a visual standard that everyone can check. For example, we might decide that those we decide to put an **X against** those in Figure 3.7 which we consider do not have beards and the others that we consider, do.

Another example might be in a restaurant. For example, a steak can rare, medium or well done. How often might you have seen a steak returned to the kitchen because it is not cooked to the liking of the customer? Obviously, there is a difference between the interpretation of the instructions by the chef to that of the customer. This same type of flaw can appy to many things.

How many whiskers make a beard?

Figure 3.7   Beards

How done is well done?

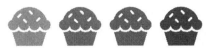

Figure 3.8   Cupcakes

Interestingly, in hospices in the UK there are visual standards for excrement to be able to record the degree by which a patient might be constipated or suffering diarrhoea! We have the illustration but spared you, dear reader, the details!

Pause for thought

Look for some opportunities where you can collect some attribute data or subjective data. In a later chapter you might decide to use it. Think about applications in your own home as well. How about a visual standard for a cup of tea or coffee so that never again will it be either too weak or too strong? If you like a medium rare steak, how about a photo of the next one that is just perfect?

Summary of learning

There are, in fact, only two different sorts of data, attribute and variable, although there is a subgroup of attribute data called subjective data because it is subjective to the human senses. Research has shown that nothing more than the bare basics tools for data gathering and analysis are necessary to deal with more than 80% (some say 90%) of all situations. Sampling always involves some risks – these can be reduced with larger sampling sizes. 100% sampling carries the grossly underestimated risk of human error.

SECTION 4: TYPES OF DATA

What you will learn in this section:

- A wide range of data gathering methods
- The errors to be expected from a variety of displays
- The importance of 'repeatability' of measurement
- Traceability.

As was stated in the last section there are basically only two kinds of data, **variable** and **attribute** (there is a subgroup of attribute data called subjective data. The visual standards shown earlier are an example of subjective attribute data, which is subject to visual inspection). Regardless of the details, all the specific data in each of these two groups can be analysed and dealt with using the same simple techniques. It is the variety of available techniques that give the impression that it is complex. Fortunately, however, as we have said, the basic tools are very simple in both cases and are perfectly adequate for most day-to-day applications.

Tools for data measurement

A common ruler is quick and simple to use, but it is far less accurate than a precision instrument. Which shall we use? Well, it depends on the situation. The latter will require considerably more skill in its use and it will be more expensive but when very precise control over a process is required and slightest changes are to be detected as soon as possible after they have occurred then we must use them. In this case we have three points to consider.

1. The skill of the operator.
2. The accuracy of the equipment in relation to the variation we need to detect. This is often referred to as gauge R&R.
3. The repeatability of the test if the measurements are taken several times in succession.
4. The degree of error as there will be errors will be due in part to both the operator and the equipment. We need to establish this with a simple using gauge R&R. We will not go into it here as it is well explained on the internet with several free software packages available that will not only explain the concept in detail but will enable you to estimate the usefulness of your measurement systems very easily.

If the tests to be carried out are expensive, time-consuming or destructive, for example, the only way to find out if a safety match does its job is to strike it. If we are a factory that produces these, nobody would thank you if you did a 100% test on all the matches in the finished goods store to see if they were OK. In this case, we will want the maximum information from the least amount of data. In this case, we will not be able to use our simple tools and seek the assistance of an expert in sampling techniques.

Fortunately, even in these cases, there are sampling tables called 'acceptance sampling' tables that we can sometimes use. But, as will be mentioned later, they are silent on the two subjects of the economic balance and human variability or error. They are simple to use but care must be taken to follow the rules. In this section, we will concentrate on the foundation techniques and just give an overview of what could be studied in the future.

Variable data

Such data includes characteristics such as dimension, weight, speed, time, voltage, current, capacitance, **viscosity, moisture content, lead times, credit period, etc.**

The type of variable data most familiar to many of us is the speedometer, rev counter, fuel and water temperature gauges in our cars. These vary within the scale getting faster–slower, full to empty, hotter–colder, etc. as we drive.

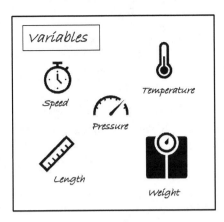

Figure 3.9   Variables

For example, in terms of accuracy, a common ruler or tape measure is quick and simple to use, but it is far less accurate than a precision instrument. Which shall we use? Well, it depends on the situation. The latter will require considerably more skill in its use and it will be more expensive but when very precise control over a process is required and slightest changes are to be detected as soon as possible after they have occurred, then we must use them. Selecting the most appropriate tool for the job is an important first step. In this case, we have three points to consider.

1.   The skill of the operator.
2.   The accuracy of the equipment in relation to the variation we need to detect.
3.   The repeatability of the test if the measurements are taken several times in succession. The degree of error – as there will be errors due in part to both the operator and the equipment. We can establish this with the simple test referred to as 'gauge R&R'.

Where accuracy of measurement is important, we find that there are also many other types of error to be avoided. Some of these relate to the equipment we intend to use. Possible equipment-related errors include:

•   Wear – this can be caused by poor maintenance or lack of calibration, excessive use, bad handling and poor storage, use by under trained personnel, etc.
•   Instability – this is common with high-precision electronic equipment and pneumatic systems where circuitry can overheat with time.
•   Operator to operator differences – causes can be lack of training, lack of care in taking measurements, lighting, task-related problems, time on task, eyesight and a host of other human related causes.
•   Environmental – dust, dirt, heat, humidity, etc. can all play a part.

Of course, the more precise the measurements need to be the more important become each of these considerations. In your work, can you think of any situations in which accuracy of measurements and errors in recording might cause problems?

Questions that you might ask yourselves:

- Are our gauges ever checked for accuracy?
- If each of us in our department all measure the same thing, do we get identical results? If not, why might that be?
- Do we get the same results at different times of the day?

Such data includes characteristics such as dimension, weight, speed, time, voltage, current, capacitance, **viscosity, moisture content, lead times, credit period, etc.** The type of variable data most familiar to many of us is the speedometer, rev counter, fuel and water temperature gauges in our cars. These vary within the scale getting faster–slower, full to empty, hotter–colder, etc. as we drive.

Attribute and subjective data

Attribute data includes all right/wrong, good/bad, is/isn't, did/didn't, error/no error, missing/not missing, etc. type situations. There is also a subgroup to this type of data called subjective data because it is subjective to the human senses, and includes activities such as wine tasting, tea tasting, loudness when measured by the ear instead of an instrument, feel, smell, etc.

Attribute data always begins from zero possible occurrences. For example, a roll of cloth may have zero, one, two, three or more oil stains, marks, or tears. In theory, it could have an infinite number. A batch of invoices might contain zero, one, two, three or more errors but it cannot have less than zero errors!

Use of charts to analyse data

Many people erroneously think that the analysis of data always requires the use of mathematics. The good news is that this is often not the case. Surprisingly, we can

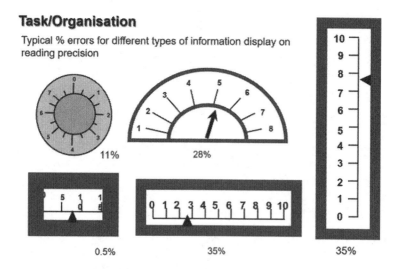

Figure 3.10   **Errors in data reading**

get much of the information we need without having to make any calculations at all. All we need is to be able to construct some very simple to use charts. Once these are prepared almost everybody can collect and analyse data.

Process measures

Process measures are those which are taken during the course of operations in order to detect unwanted deviations from the plan. In a manufacturing situation this might involve measurements taken by hand with gauges of various types such as clock gauges, micrometres, rulers, tape measures, air and electronic gauges, coordinate measuring machines, and both assisted (magnifiers) or unassisted visual inspection. It also includes a wide range of non-destructive testing methods such as magnetic ink detection, fluorescent die penetrants, echo scanning in various forms and a range of destructive tests.

Researchers claim that over 80% of all inspection is by the human eye using no aids of any kind. In many situations the measurements might also be automatic or semi-automatic. This can include direct measurements involving mechanical devices, inductance, capacitance, electrical current, voltage and resistance, high speed electronic optical systems using cameras, air gauging etc.

In most cases other than where ratios of wavelengths of light are used or where one set of data is compared with another from the same measuring source, calibration and traceability of measurement is important, especially where high precision is a requirement.

In non-manufacturing, all the same principles apply as to manufacturing. It is just that the tools may sometimes be different.

In fact, we can find monitoring and measuring equipment all around us in all walks of life. Probably one of the most universal these days is bar coding. This is used in manufacturing for defect coding, warehousing for stock control and, of course, this is also used in the retail industry extensively.

There are many fewer tools for measuring attribute data than for measuring variable data because much of it relies on human observation but there is one important group that are referred to as go/no go gauges. For example, for deciding if a bored hole is the right size.

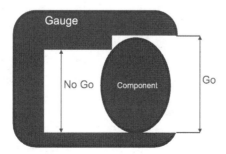

Figure 3.11   Go/no go gauge

Typical tools for measuring attribute data

It can be seen from Figure 3.11 that the means of measuring for attribute data can be relatively simple, and less skill is required to collect the data than for variable data. But as usual there is a trade off! Whilst, in general, tools for measuring as an attribute are simpler and quicker to use, as a rule you need around ten times as much data to have a particular knowledge of the state of a process than might have been the case if variables type measurements had been taken.

Other forms of attribute sampling.

There are metal detectors at airports, shop exits, etc. There are all the dials and gauges that light up as soon as we switch on the ignition of our cars. There, we are confronted with a range of gauges and sensor lights as is the case with aircraft, ships and boats and, more recently, GPS and satellite navigation. All of these can be regarded as forms of in process control. Many require calibration, and most are subject to error both within themselves and especially at the gauge/human interface.

Pause for thought

We have covered more ground in this section, so it might be an idea to look back. We mentioned gauge accuracy and the fallibility of the human inspector. If it turns out that it is not as good as you might have thought, do not blame the inspector, the problem is almost certainly job design. Have you any idea of the actual performance of your inspectors, their equipment, its maintenance and therefore traceability of measurement? It is something to think about for most. We do not have the space to go into it here, but you might google vigilance decrement, time on task, multiple monitors, signal-to-noise ratio, Mackworth clock test, Broadbent's single channel theory, and a myriad of scientific papers on this subject.

Summary of learning

You have learned that the range of different data recording techniques is large and that all of them are subject to error of one sort or another, but these errors can be either avoided or reduced.

You also learned that, especially with precise forms of measurement, there is variation in results due either to the equipment or the person taking the measurements.

Because machines can wear or drift, it is important to recalibrate frequently using gauges that have themselves been checked to International standards.

SECTION 5: USE OF CHARTS TO ANALYSE DATA

What you will learn in this section:

- The value of charts as a way of interpreting data.

Only the most popular charts are referred to here because there are hundreds of variants to most of these and new ones are being created all the time. At almost every election there is an elaborate variation on the concept of a 'swingometer' on TV and each major channel will have its own unique version.

It is also useful to have knowledge of a variety of chart types because, in a given situation, one type might tell us very little; reorganise the data and use a different type and it might produce stunning results. This happens frequently.

It is likely that everyone is familiar with the sort of chart in Figure 3.12. Sometimes, worryingly, it is bleeping away over ours or a loved one's bed. More often fortunately, we are viewing it on a hospital soap on TV. We may not be able to interpret every little ripple on the line, but we all have some idea what it means. We sure know to panic if it suddenly becomes a steady straight line with a continuous bleep.

Sometimes, we do not actually record data but often we do, for example, the data on this heart rate monitor. In some sophisticated models not only will it give us an instantaneous read-out on the screen, but its memory can store the data and we can present it back to ourselves in several ways depending on what we want to know.

For example, whilst we are training, we are only interested in the heart rate at that time and the single number is all we want. If you are an athlete, it is a good idea to print off or electronically store the data to make time-to-time comparisons

After the ride or run, we may want to analyse the results and see how the rate varied over a period, so, we might study the frequency of the beats shown on the chart on the right in Figure 3.??. If we spot any irregularities, we might take the data along to our doctors for more expert advice.

Alternatively, we might want to analyse the data in yet another way. For example, we might want to analyse the amount of time in the training session that our heart was at a level. We can see here if we look carefully at Figure 3.???, that the

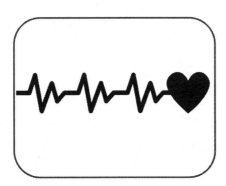

Figure 3.12   **Heart rate monitor**

user spent most of his/her four hours training time working out at 164 bpm but with some activity much higher.

The current heart rate shows at 72 and the resting pulse is 56. Probably, most of us can interpret this sort of data when it interests us enough and when it is as simple as strapping a band round the chest and a monitor on our wrist. However, even if we did not have the means to do this electronically, to do it manually is really no more difficult. We just have to apply some simple rules.

Why use charts?

We use charts because it is easy to understand them, whereas looking at columns of numbers can be confusing and even if we took the time to study the data, we could easily make mistakes.

For example, look at the table of data in Figure 3.13, which records the number of miles travelled on a series of tanks of petrol. What does it tell us? Well not very much. A part from having a rough idea that the vehicle does somewhere around 300 miles per tankful, very little.

Try recording this for your car if you own one. I suppose you could say that the usage varies quite a bit from tankful to tankful but how much does it vary? Is the variation getting greater or smaller? This is not at all obvious just by looking at a matrix of data

If we look hard enough, we can see that there is one reading as low as 288 miles but at the other extreme there is one at 362. Is this reasonable variation or not? Well, we would not expect the data to be identical. We would be very suspicious if it were. But just by looking at the data we can learn very little so let's put it in chart form and see what happens then.

Well, that is a bit different. All we did was to get some squared paper, or we could have done it using a computer programme such as Excel if we knew how. If we plot each piece of data in sequence let's see what it has revealed.

First, we can see the expected tank-to-tank variation (Figure 3.14). We can also see if we look carefully at the trend line that there is a gradual downhill drift. In fact, it has on average dropped from approximately 345 miles per tankful to around 320. There is no way we could have seen that from just looking at the numbers. But why has it dropped? If we had not charted the data, we probably would not even have known we had a problem! The chart is doing its job!

Before we move on, here is something else the chart tells us. Yes, there is variation from tank to tank but just look at the size of the variation in the ringed area of the chart (Figure 3.15). This looks abnormal compared with the rest. Maybe there is a reason for it?

342, 332, 290, 301, 333, 319, 342, 288, 344,
316, 351, 340, 315, 319, 299, 341, 354, 300,
290, 296, 355, 349, 360, 316, 322, 338, 316,
350, 318, 362,
351, 340, 338, 329, 344, 354, 322, 347, 360,
352, 328, 320

Figure 3.13   Data table

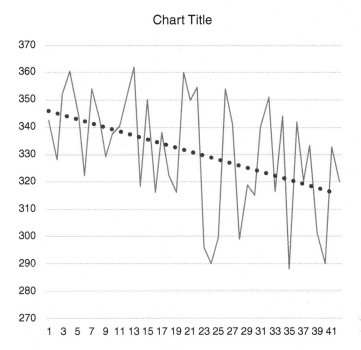

Figure 3.14   Charted data

When we dig in and if we also recorded the journeys we made during that period (we might get it from our diaries). Maybe we discover that our kids borrowed the car during that time, or maybe we went on holiday to the Lake District and used the car to climb a lot of hills. There is so much possible information from just one simple table of data. Figure 3.16 is the same data but, in this case, presented as a frequency diagram or histogram.

The choice of chart depends upon what you want to show. The previous chart (Figure 3.15) showed the time-to-time trend. Figure 3.16 gives a graphic visualisation of the spread of the data. This is similar to the chart shown on the heart rate monitor earlier (Figure 3.12).

Do you remember what 'attribute data' is? Everyone has a struggle with these strange words at first, but you will soon get used to them. Anyway, if you do not want to go back, briefly, attribute data is always:

- Good/bad
- Is/is not
- Right/wrong/
- Taste's nice does not taste nice, etc.

Attribute data always begins from zero possible occurrences. For example, a roll of cloth may have zero, one, two, three or more oil stains, marks, or tears. In theory it could have an infinite number. A batch of invoices might contain zero, one, two, three or more errors but it cannot have less than zero errors! You cannot be struck by lightning less than zero times! You might have 0, 1, 2, 3 more emails in a day

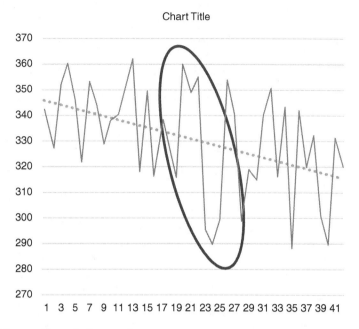

Figure 3.15   Chart information

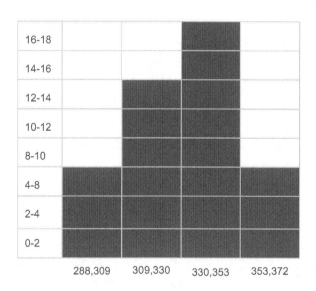

Figure 3.16   Frequency diagram

possibly hundreds but the number is attribute data and if you decided to plot the data on a chart you could use the same techniques as for the others, etc. For example, Figure 3.17 show a process with a stable 7% defects.

Note that with attribute data when the mean is less than roughly 15% it will be seen that the spread of the data leans very slightly to the left and the closer to zero it becomes, then the more skewed it will be. But with the average or mean being 7% or increasingly further from the left axis, it will appear to be ever more or less evenly distributed about the mean. In fact, it will appear like a normal distribution curve for variable data. As the calculations and tables for this are easier to use, then in certain circumstances, it is acceptable to make that assumption, but you do need to be careful and check with a specialist first if you do not feel confident. Now, supposing an improvement had been made to the process and data again plotted to see the effect of the changes. Figure 3.18 shows what has happened.

Note that as we have said, the average or mean is moved to the left because of an improvement, the chart becomes skewed with a short tail on the left and a larger one on the right. If there is only one significant variable, this will generally be the case and is positive proof that something has improved

Now, let us look at the situation after yet another improvement. The result might well look like this. Instead of the chart being skewed as the average number of defects approaches zero, the chart will appear increasingly one sided. This is because, of course, it is impossible to have fewer than zero defects.

Case 1: the leaky catheter bags

Let us take a look at a real-life situation. Wearing a leaky catheter bag is not a popular idea so it is hardly surprising that the company, which makes this product conducts

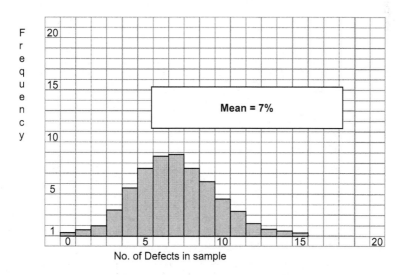

Figure 3.17   Variation of data – large spread

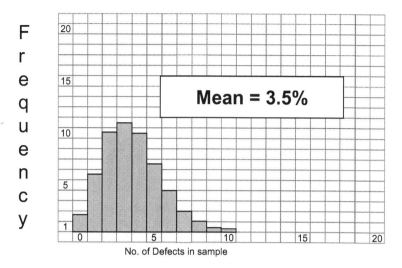

Figure 3.18   Medium spread

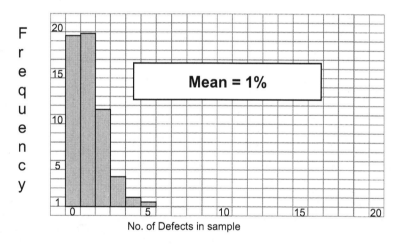

Figure 3.19   Small spread

leak tests on 100% of its products. The test requires that the product is immersed in a tank of water to a depth of approximately 5½ inches. Air is then pumped into the bag at a pressure of 7 psi; bubbles indicate any leaks. Until we conducted a training course for the production staff, no one had ever analysed the data!

Two of our students then took 100 batch cards from the production process (each card contained the results of the tests on 100 products which was the batch size) and charted the number of defects per 100 batches.

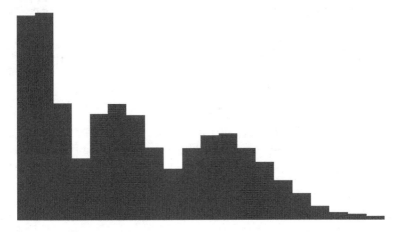

Figure 3.20   **Catheter bag data**

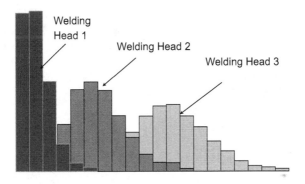

Figure 3.21   **Stratification**

One of the team members then suggested the possibility that the chart might be concealing three separate sets of data. The question then arose, what might cause the difference? One member then remarked that there were three different welding heads: one in the store, one on the machine and the other being cleaned. Perhaps there was a head-to-head difference. When the data was re-plotted for each head separately, this was found to be the case:

The challenge for the team was then to find out why one of the heads was better than the other two. When they found the answer, they were able to bring these up to the level of the best. The total time spent on the data collection and analysis was three hours. The cost of the improvement was negligible. The benefit saved the company thousands of pounds per annum in defect costs, not to mention the improved reputation with its customer the Department of Health and Social Security.

The technique was extremely simple, and the problem solved by a team of process workers! Almost anyone in the factory could have carried out that analysis at any time over the previous 25 years. Why did they not do it? The answer is simple. Until our training course, nobody in the factory knew how to analyse the data. Well, maybe somebody could have done it but whoever it was probably never went down into the factory, did not understand the processes and, in any case, it was not their job!

Analysing variable data

Mathematically or theoretically, the treatment of variable data would appear to be quite different from attribute data. This is true, but fortunately the graphical techniques and the reasoning when analysing the results is remarkably similar. In the use of the graphical techniques, whilst the shape of the distributions of date may in some case be slightly different, the ability to see what is going on is just the same.

The charts that we used in the attribute sampling above are referred to as frequency diagrams or (in a special case, tally checks. For practical purposes, in project work and for very simple control purposes, the means of plotting the data is very similar to the plotting of attribute data with the only minor difficulty being the selection of scales on the graph. However, this variation on the 'frequency diagram graph' is usually referred to as a 'histogram'.

Both the histogram and the frequency diagrams are forms of 'column graphs' but with one important difference from all other forms of column graph. In the case of these two, the data is continuous, and each column is directly connected to others on its left and right. We cannot take any of these columns and move them around as we can in another type of column graph called a Pareto chart. A histogram is created by recording data graphically in columns where each column represents a tight cluster of data within predetermined limits. For example, we might make the columns for recording the temperature of a room as follows: Each time we take a measurement we put an X in the corresponding column as shown

If temperatures are recorded over a period, the result might look like this: where we draw a box to include each column of Xs. Usually but not always, it will look like Figure 3.22.

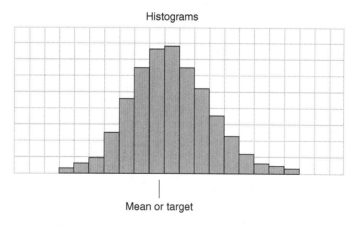

Figure 3.22   **Histogram**

The cost of variation

As mentioned in Chapter 2, it was Dr Taguchi, the famous Japanese engineer/ statistician who in the late 1950s first demonstrated using his famous '**quality-related costs are a quadratic function**' argument that quality-related costs go up as variation increases, independently of any arbitrary limits set by designers. This means that variation is always a cost. We will not worry about it here, but he was able to prove that variation cost money and represented a loss to society.

For example, a dustbin lid would have quite a large allowance given on the diameter because if it went over the bin itself it would not appear to matter much if it was a sloppy fit. But the bigger lid the more material is used and if it is smaller than the bin then it will not fit!

On the other hand, a precision component in a pressure valve on a space vehicle on a mission to Saturn or similar would need to be as near perfect as it was possible to be. Too much variation here could destroy the entire mission. For this reason, all dimensions are given tolerances. In many cases, the designer was not sure of the effects of large tolerances, but he knew that small tolerances were safer from a function point of view. For this reason, he would err on the safe side and many tolerances were tighter than probably necessary. If we can discover the causes of variation and remove them it is better all round!

Figure 3.23   Dustbin

Pause for thought

Focusing just on attribute data situations, make a note of a few where these simple charts might be used and try out the method. You can do this at home and in your pastimes, too, especially in sport. The opportunities are myriad. How about the number of goals scored by your favourite team home and away? Does it change when there is a change of manager or a key striker? Maybe it could help win a few bets on the horses!

Summary of learning

You have learned most data can be presented in chart form and that different types of chart may reveal different facts. This was illustrated in the miles per tankful example (see Figure 3.??).

You saw how there are many types of chart. We covered those that are used an estimated 80% of the time and which will serve your needs in most instances. You also saw the late Dr Taguchi's observation that all variation is a loss to society, so we should always be on the lookout for ways to reduce it. This is covered in some detain in the next chapter.

SECTION 6: THE STUDY OF VARIATION USING CONTROL CHARTS

In this section you will learn:

• The use of control charts for studying and controlling variation.

It is a simple step to go from the static charts that we have been looking at in Section 5 to the use of active charts. These move with the data to let us see what is happening in our processes, when they happen and to enable us to make corrections or observe trends as they happen.

Control charts can range from being very simple to use or they may be quite sophisticated for special cases. Fortunately, for most applications we do not have to go deep into mathematics.

This is just an overview of the use of control charts for studying and controlling variation. The following figures illustrate how control charts can be used to identify different types of special causes. A control chart can be constructed on standard graph paper, but it is more usual to use a form specially printed for that purpose. The following chart (Figure 3.24) shows variation on the vertical or Y axis of the chart and time along the horizontal or X axis. The first reading, which is on the left, shows that the data is between the two limits upper control limit (UCL) and the lower control limit (LCL) but it is fairly near the upper limit. The second reading is similar, then there appears to be a trend downwards. This can happen purely by chance, but it could also be an indicator of change.

Control Charts

Upward and Downward Pattern

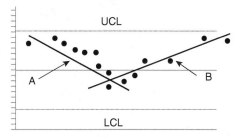

Patterns A and B each indicate an out of control
process, they have 7 or more successive points
in a decreasing/increasing direction.

Figure 3.24   Control chart trends

Static Patterns

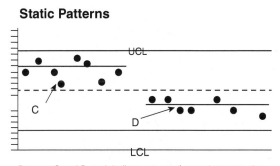

Patterns C and D each indicate an out of control process, they
have seven successive points above/below the mean.

Figure 3.25   Control chart step change

Convention tells us that if we get a trend of seven or more readings in the
same direction then the process has probably become unstable and the mean is
drifting, in this case downwards. The next seven readings indicate a trend in the
opposite direction. This is a simple reading of the chart, but it does show that
such charts can make a dramatic improvement to quality if used properly.

Typically, these data patterns indicate either a downward or upward drift.
This is often caused by worn tooling, a gradual slackening of adjustment, overheating etc. If there are fewer than seven consecutive data in the same direction,
it is unwise to make such an assumption because it might just be random run in
the data. At seven or more points, that is still theoretically possible, but the likelihood will be less than one in a thousand occurrences of samples if four or more
are taken.

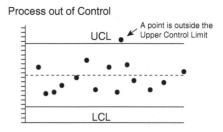

Figure 3.26   **Out of control on limits**

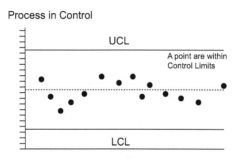

Figure 3.27   **Control chart – in control**

These data patterns indicate a possible step change rather than drift. It could be due to the use of a new batch of raw material or a jump in the process settings either for internal or external reasons.

Figure 3.26 indicates that the process has become less stable than when the chart was originally constructed. It could be due to a lack of homogeneity in the materials, a slackening of controls, worn tools etc.

The chart in Figure 3.26 is typical of a process that is well in control about the upper and lower control lines. The variation is reasonably symmetrical around the mean and the spread much as might be expected within the control lines.

The introduction of basic statistical methods enabled a more affective diagnosis of the causes of problems and to begin to drive out variability. Eventually, many quality circles teams progressed to more sophisticated statistical tools. This reduction in variation led to a reduction in unpredictability, which in turn led to smoother production performance, lower levels of inventory, just in time and all the methods of lean manufacturing. All of this happened in the late 1960s and early 1970s.

If you have quality circles in your organisation you could ask for training in this important technique. If you do not have quality circles maybe you could ask if they could be set up. People in quality circles say that it has transformed their work from drudgery to being something they look forward to.

Examples of reducing the cost of variation

Case 1: service time – counter staff

When the average service time spent dealing with customers in a bank, shop, check in at a hotel, or other similar data is plotted, the information revealed can determine the number of staff required to meet a given level of demands. It can also indicate the dispersion transactions for different types of service and point to possible means for improvement.

For example, it is possible that the dispersion may be due to differences in types of customer demands. By segregating these different customers, the service time for some may be greatly reduced. It can be seen from the foregoing example that much can be revealed from the simplest of techniques – techniques that can be used by anyone, without any need for mathematics or deep statistical theory. Techniques such as those described above are taught in simple but effective educational programmes aimed to empower the workforce. Through training in these basic skills, Japanese workers have managed to solve literally millions of problems of the level described in this text. Before moving on, let us look at a further example of the power of these simple techniques.

Case 2: bandage weaving

The earlier example of the catheter bags was recorded during a real-life example obtained from the first day of a three-day course in basic data collection and analysis for line managers and supervisors. None had any previous educational background, and in just a short time, thousands of pounds of real savings had been made using the tools described.

Figure 3.28   Bandage packing

On the second day, the participants were introduced to data collection using variable data. One group decided to check the lengths of bandages ready for shipment in the finished goods stores. Printed on the boxes, the length was specified as being 10 metres. About 50 boxes were selected at random and lengths measured. The results were as shown in Figure 3.29.

Figure 3.29 shows that, on average, the company was giving away over 1 metre of bandage per box. It was estimated that this had probably been going on for some 30 years and cost the company £35,000 for each separate size of bandage. For a small company, this was a great deal of money. The worst aspect was the fact that the customer was probably totally unaware of the gift. How many people measure bandages? Certainly, no one other than a trading standards officer would even bother.

Case 3: plant maintenance

Using the same technique as described above, an engineering company obtained the co-operation of its workers to collect data on plant breakdowns. Variable data were plotted for time to failure on certain components, the components known as 'feed fingers' – metal parts used to push the product into the machine, which were known to have relatively short lives.

If failures occurred during the shift, disruption was high and cost around two hours of downtime for each occurrence. It was found that, on average, this occurred about every 260 hours of operation. This did not seem to be a big problem but in fact the real

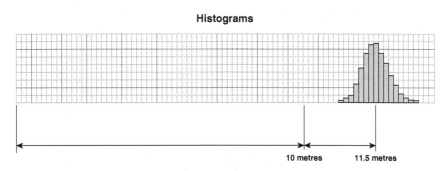

Figure 3.29   Bandage data – length

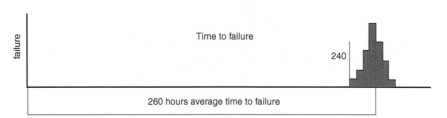

Figure 3.30   Time to failure data

cost was much higher than was realised, not in the direct cost of lost time, but in the accumulation of work in progress (most of it is not progressing). There was also the question of reduced predictability of production output, which, in turn, led to stock outages, and to keeping excessive stocks in finished goods warehouses to safeguard against this. These costs appear on the balance sheet as stocks and inventory and also suffer depreciation and incur interest charges. Worse still, these unscheduled break-downs result in lower levels of plant utilisation. The usual counter to this would be to obtain additional plant or machinery/equipment, and of course, set appropriate man-ning levels. These will appear as direct costs on the profit and loss account, fixed but depreciating assets on the balance sheet will tie up precious working capital.

When the failure pattern was revealed, it was agreed that the feed fingers should be replaced every 240 hours even though they were still functional. These changes were made outside normal shift time, therefore causing no disruption of work flow, better pre-dictability of scheduling times, no log jams in production and lower work in progress. This activity is known as 'preventive maintenance' and in relevant situations can save huge sums of money in wasted time. When collected by the workforce, this data is almost free, and the workers enjoy being involved. Breakdowns are probably as much a frustration for them as they are to the scheduling department, and to the managing dir-ector when he is forced to analyse his factory costs. In Japan, it is estimated that over 80% of all problem solving requires nothing more than the application of these simple tools.

Case Example 4: 'the cracked bed!'

A medium-sized company employing around 400 people in the automotive compo-nents industry experimented with the implementation of Xbar/R control charts in the manufacturing of high-precision automotive gudgeon pins. Nevertheless, whilst

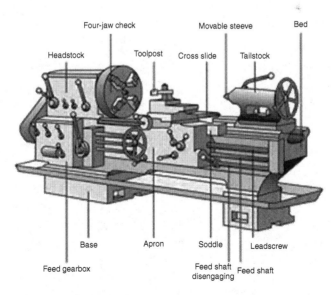

Figure 3.31   Process capability study

these demands required the very highest levels of machine capability at the final stages of production, surprisingly wide latitudes of dimensional accuracy were tolerated in the early production processes. These inaccuracies were thought to be unimportant prior to the heat treatment process as it was thought that, at this stage, scale and other surface problems were such that considerable grinding would be necessary afterwards at which time the variations could be removed.

The first experiment was conducted on a 3-inch (maximum diameter of the bar stock that could be used) automatic lathe. It was found that several of the readings were outside the statistical control limit lines, which had been calculated from the data, and this did not seem to make sense. The question was, why?

At first, it was thought that some error had been made in the calculations and it was checked several times, as was the theory and the measurements of the components. The engineer making the study consulted the setters of the machines, and also the operators. Many possible theories were suggested as to why this should be the case. The suggestion by one setter – 'cracked bed' proved correct. The machine had been purchased second hand and apparently it was known to the setters that the machine bed was cracked from the time of purchase. When asked why they hadn't volunteered this before, the reply was 'no one ever asked!'

The bed of the machine was repaired by the tool room. Again, samples were taken and still the results behaved in an abnormal way. However, whilst the engineer was in attendance at the machine, he noticed that one of the operators checked a component and decided to adjust the size – he did this with the aid of a mallet.

Several other theories were also tested and some of these also proved correct. The resulting improvements drove the variation down and down. Eventually, if they went further the cure would cost more than the study so the engineers went on to study other machines and make further improvements. Hundreds of thousands of pounds was eventually saved.

Pause for thought

Are these simple chart techniques used in your organisation? Can you see opportunities for them? If so, you might need external assistance in getting people trained. You need to be wary. The market is saturated with those who would welcome the opportunity to relieve you of money but be cautious. There are two things we would recommend. First, avoid anyone who offers to use software for this purpose. It is not that it does not work or that it is not quick but, from our experience, people need to get their sleeves rolled up, draw the charts just as shown here and study the results. If the charts do not reveal what the team are looking for, then vary the techniques. Not only will they get to understand the tools they are using better, but they will get a better understanding of the processes, too. Second, always demand experiential learning on the job. All the examples we have used here were on the first morning of the first day of training and already tens of thousands of pounds were saved, the biggest being the rolls of bandages example. Those who do all of their teaching in the classroom are unlikely to be able help produce such results. Finally, before going outside, just check. Is there nobody on your own payroll who can do it? Often there is, and they will probably love the opportunity. Also, they should be more or less permanently accessible. We are not talking about rocket science here.

SECTION 7: USING DATA FOR INSPECTION

In this section you will learn:

- How sampling can be used
- The risks of sampling
- Operating characteristic curves
- Economic sampling plans
- The reliability of human inspection
- Inspection failures.

Introduction to sampling plans for acceptance testing

So far, we have looked at the use of statistical charting methods for the analysis of variable and attribute data. We now need to look at how sampling can be used not only for analysis but for deciding whether a given batch of product is likely to be acceptable or not without having to do 100% checking. In many cases, 100% inspection will not be economical and in other cases it might not even be practical, as in the final inspection of products such as matches, fireworks, missiles, fire extinguishers etc. In these cases, we need to be able to take the smallest sample possible to make the best decision possible. For this purpose, we can make use of British Standard 6001 acceptance sampling tables.

Acceptance sampling plans

These take all the drudgery out of designing your own sampling plans, which nobody does. However, there is a trap into which most people fall when using these tables and unfortunately the tables are not helpful in its avoidance. Before we explain that, let us look at the basis of the tables. As we have said earlier, sampling takes a risk. When we sample, we take two risks.

- Type 1 is believing that the data says something is bad when it is not
- Type 2 is believing that the data says something is good when it is not.

These two risks are always present in sampling. What we have to do is to keep them to the minimum. One way to do this is to take a larger sample. **Theoretically, the larger the sample the smaller these two risks become, and they theoretically reach zero when we reach the 100% sample level.**

We say 'theoretically', because this overlooks another important fact. Human beings are vulnerable. 100% inspection does **not** achieve 100% segregation of good from bad. It would be nice if it did, but it does not. Not only do people make mistakes, but the number they are likely to make varies substantially from one situation to another. Numerous scientific studies have shown that in some cases it is not unusual for humans to miss as many as 90% of the errors! Fortunately, help is at hand to make things easier.

There are published sampling plans to take the drudgery out of calculating sampling sizes based on risk, but they are not perfect because they do not take human factors into account. They are however the best we have.

In the UK they are known as BS 6001 acceptance sampling tables. These are based on some American Standards Mil Standard 105D. They give a collection of graphs known as operating characteristic curves from which sampling risks can be determined and from associated tables. Your job might not involve you in this subject, but the ideas are worth bearing in mind when thinking about taking samples.

We are only covering the subject briefly here but If you do need to know more, the originators, US Military Standards under Mil Standard 105E have published some very clear notes, which are printed in the Standard and these are reproduced in the equivalent British Standard sampling plans.

This chart in Figure 3.32, known as an 'operating characteristic curve', shows a given sampling plan – a sample of 80 items from a batch. If there is a reject in the sample, we reject the batch and if there is not, then we accept it. This is why it is often referred to as 'acceptance sampling'.

Note that as the average number of errors in the batch increases, then the likelihood of rejection increases as we would expect, as it is more likely that one of the rejects will appear in our sample.

Notice also that at zero errors the probability of acceptance is '1'. In other words, 100% certain. Notice also, that as the number of errors increases, the probability of acceptance decreases as you would expect. However, it does so at a varying rate.

It only drops slowly at first then it drops fast, before slowing down again later. This is typical for sampling plans and the shape varies depending on the size of sample taken. As we said, this one was calculated for a sample of 80 items. They could just as easily have been sweets as missiles, the chart would have looked the same.

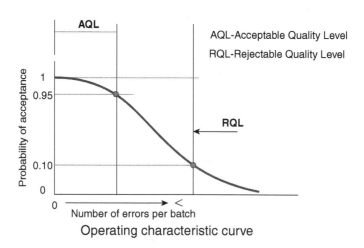

Operating characteristic curve

Figure 3.32   Operating characteristic curve

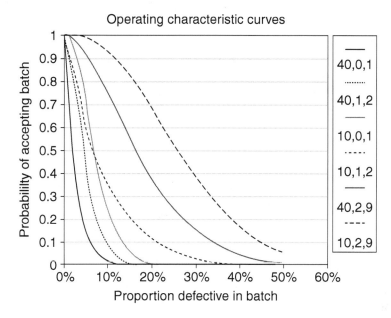

Figure 3.33   Multiple nested OC curves

Note that at one point there is a .95 (or 95%) chance of acceptance. This is known as the AQL or acceptable quality level and is usually a contractual figure between a supplier and his customer.

Note also that as the number of defects continues to increase, it reaches a point where only 0.10 (or 10%) will be accepted and this is known as the rejectable quality level. Mathematically, there is only one possible sampling plan that will pass exactly through any two pre-selected points.

To save having to do the calculations there is a published set of tables with hundreds of sampling plans to suit a wide range of needs. These plans include a large number of these graphs and, typically, the user finds one that is closest to his or her needs and then chooses that plan from the tables. An illustration of these nested curves is shown in Figure 3.33. These plans are simple to use but care must be taken to follow the rules. This is as far as we will cover this topic in this unit as it is a general overview but, if it is useful to you, there is a lot of explanatory detail in the Standards themselves.

In the USA these curves are contained in a Standard known as Mil Standard 105E and in the UK BS 6001. Since the contents are identical, if you do wish to use them it would be an idea to do a cost comparison.

For those who wish to use the BS6001 plans, be wary of two points. First, it will be noted in the tables that the sample sizes for a given AQL vary according to the size of the sample. This is statistically invalid because the operating characteristic curve is independent of the size of the batch.

The reason that they have done it (but it does not say so) is that the tables assume that people would prefer a lower level of risk for larger batch sizes. This may be true, but we think it is something that the user should decide not the tables. Again, refer to the curves.

The second point is that the tables are based on a series of AQLs, say 1%, etc. However, they are silent as to how to choose the most appropriate AQL. This should be an economic decision and considerations include the consequential cost of a defective part getting through, the cost of doing the inspection per part and for the scheme being considered.

Pause for thought

Think about situations where you or others take random samples just by taking a handful or flicking through some files. Is the data valid or is the resulting opinion risky?

What might be done instead?

SECTION 8: INSPECTION

In this section you will learn:

1. The reliability of human inspection.
2. Inspection failures.
3. Workplace design.

Man vs machine

The word 'inspection' can apply to visual inspection by a human being but it also can be by machine. Each has its advantages and pitfalls as shown in the following examples.

Jobs where people are often better than machines (this is changing all the time as machines become more and more intelligent):

- Detection of signal in noise
- Pattern recognition
- Adaptive – learning situations
- Good for subjective classifications
- Flexible – can cope with product changes
- Can assess a wide variety of fault types
- Process knowledge can be used
- Can move to the work.

As stated, this is rapidly changing as computing power increases.

Typical disadvantages of machines (at present!):

- Cost of initial set-up
- Need for maintenance (calibration although in different ways this is also true for humans)

- Lack of flexibility
- Only standardised feedback available
- Obsolescence
- Wear out
- Work must generally be presented to the machine.

As stated, this is rapidly changing as computing power increases

Inspection failures

1975 Chapman and Sinclair

Defective tarts in tart manufacture:

37% missed on fast conveyor + 2.5% false alarms
*15% missed on slow conveyor + 1.5% false alarms.

Defective chickens on packaging line:

14% missed on fast conveyor
*29% missed on slow conveyor!

1972 Schoonard et al.

Electronic chip defects in manufacture:

23% no prior knowledge of likely defects
17% with prior knowledge.

1969 Yerushalmin XRay study – chest X-ray

TB detection using particularly well-qualified radiographers!

Patients thought to be suspect – 26.9% missed
Patients thought to be urgent – 25.4% missed.
Patients thought to be active – 25% missed.

Overall, there were over 20% indications present in the 15,000 plates that were missed.
Reference – detection of signal in noise.

1941 Tiffin and Rogers – paint spray defects

Surface blemishes – 41% missed
Unevenness – 24% missed
Coating thickness – 65% missed

1950 Hayes – piston ring surface defects

33% missed

1969 Drury and Sheehan – hooks

19% when no forward feed of defect type
8% if previous knowledge of type of defect

1932 Wyatt & Langdon – metal box examination

30% defects overlooked

1945 Lawshe & Tiffing – gauge and instrument reading

20% error using gauge
91% error – calliper and micrometer combined

Physical aspects

Colour blindness (or weakness)

Where colour matching is part of a process, colour tests should be carried out – remember, about 1 in 10 males have some colour deficiency.

Human variation – possible factors

- Sex
- Eyesight
- Lifestyle
- Intelligence
- Personality – introvert/extrovert
- Age
- Motivation
- Training.

Workplace design – lighting

Avoid glare. Set lights so that they do not reflect back from the work directly into the eyes.

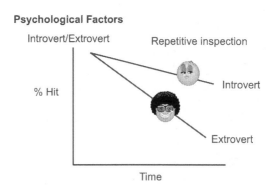

Figure 3.34   Introvert/extrovert

When new, fluorescent lights flicker at 50 cycles and it is not visible. When old, this reduces and becomes visible at about 30 cycles and causes eye fatigue. The following array is confusing, and it is difficult to see if any of the gauges are showing a wrong reading

Failure to detect error – part of report on an East Midlands plane crash

* Coincidence – The smoke disappeared after shutting down the right engine and the vibrations lessened. This is an example of 'confirmation bias'.
* Lapse in procedure – After shutting down the right engine the pilot began checking all meters and reviewing decisions but stopped after being interrupted by a transmission from the airport asking him to descend to 12,000 ft.
* Lack of communication – The cabin crew and passengers could see the left engine was on fire, but did not inform the pilot, even when the pilot announced he was shutting down the incorrect engine.
* Design issue – The vibration meters would have shown a problem with the left engine but were too difficult to read. There was no alarm.

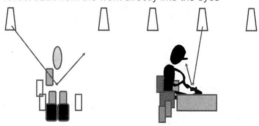

Figure 3.35   Effect of lighting

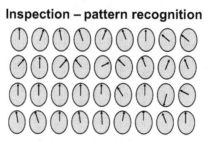

Figure 3.36   Pattern recognition – dials

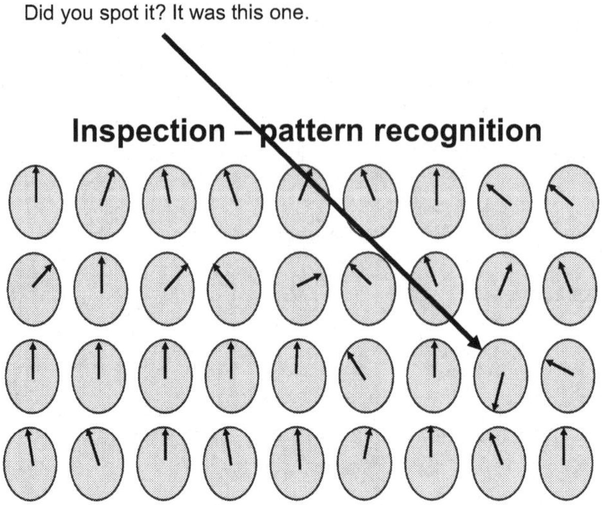

Did you spot it? It was this one.

Inspection – pattern recognition

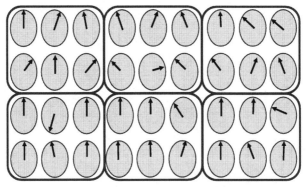

Figure 3.37   **Pattern recognition – finding the flaw**

This design is better. It uses the principle of 'enclosure'

Inspection – enclosure

Figure 3.38   **Pattern recognition – effect of 'enclosure'**

Pause for thought

Why not check the inspection methods used in your organisation, the dials and gauges, the lighting, strip lights, are gauges calibrated? What do you know about inspector error?

But it is even more easy to spot an error with this design

Inspection – enclosure

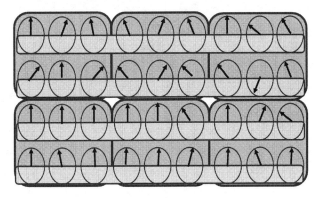

Figure 3.39   **Effect of highlighting errors**

Scale Divisions

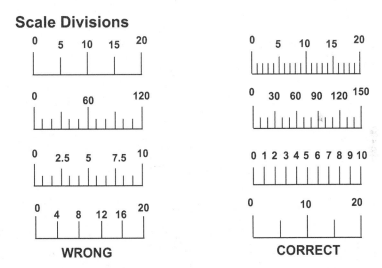

WRONG   CORRECT

Sub divisions should correspond to values of 5 – 2 – or 1

Figure 3.40   **Dial reading errors**

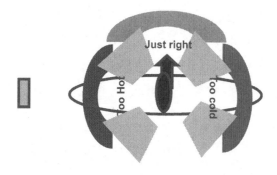

A display Instrument must show the information required
clearly and simply

There is no need to include unnecessary clutter that can
lead to errors

Figure 3.41   Dial simplification

Summary of learning

In this section you will have learned:

- The word 'inspection' can apply to visual inspection by a human being but also it can be by machine, each has its advantages and pitfalls.
- To err is to be human. However hard we try, we miss things, make mistakes, forget things and we all do it. The challenge is to reduce human error as much as possible, but we can never eliminate it.

SECTION 9: KEY PERFORMANCE INDICATORS

In this section you will learn:

- If we cannot measure it, we cannot manage it
- What are performance Indicators?
- What are key performance indicators?

The purpose of key performance indicators is to find those measures of performance and targets and goals that we know will enable us to achieve our vision. There will be more on this in Chapter 5. Of course, there might be hundreds of different measures that relate to our goal, but these will not all be equally important, and we can refer to them as being simply 'performance indicators'. Only a few of them are likely to be 'critical' and unless we set targets or goals for these then we have no chance of achieving our aim. If we can identify them, these critical goals can be highlighted and separated from the rest by calling them key performance indicators. Sometimes, these are imposed and sometimes they are self-generated and are disciplines that we make voluntarily.

An example of a simple voluntary key performance Indicator might be:

> *I want to look smart and I have decided that to do so, I will either shave every day or trim my beard regularly, wear a new-looking jacket and tie, clean my shoes, brush my hair, etc. Collectively, these small performance indicators will, if we get it right, result in looking smart.*

So, we can set key performance indicators for just about anything.

Basically, we start with a vision. What is my image of successful outcome? This could be a children's birthday party, your next holiday, success in a sporting event anything. Try to think of some of your own, it can also be the vision of your organisation.

Organisation vision

Imagine that you are contemplating starting your own business. Let us say it is a pub. What do you have in mind? A small country pub in a village? A small group of pubs owned by you and each with a manager? Or at the other extreme are you thinking big and have an idea to go head to head with one of the giants? It could be any of these and more.

OK, let's say you have in mind a country pub. You find one on the market. OK, maybe it needs a lot of work doing to it. It has been shut down for a while and lost all its customers. You decide that you will take it on. Now comes the image, several images. First, you must clean it up and make it look attractive just to get some people in the door. But having done that, you want them not only to come back but to spread the word that you are doing a great job. Here comes the first key image. What does success look like?

- How many customers are you aiming for?
- What monthly turnover?
- What services will you provide?
- How many staff do you think you will need?
- What sort of service do you want them to give?
- How will you know when they are properly trained?

We can see therefore that these can either be 'within process', i.e. at the outputs and inputs between process elements, or at the output of a process.

'Within process' KPIs are often more important than those at the output stage because if these are not met during operations then provided they have been correctly selected, then the end of process output KPIs will also not be achieved.

There are two kinds of end of process KPI. Those that are imposed by some outside interest and those internally selected within the organisation itself.

An example of externally imposed KPIs is the requirement that more than 75% of the time, paramedics must actually be at the scene of an emergency within 9 minutes of the receipt of the call at the call centre. Another is the requirement that the time on hospital waiting lists must be less than six months duration. An example of an internally imposed KPI might be that 98% of all orders must be delivered within 24 hours.

Surveys to obtain KPIs

Surveys are yet another form of monitoring and measuring. Most of us at some time have been asked to participate in market research surveys either online or whilst waiting for an aircraft in the departure lounge. These usually involve opinion data, which is usually the most difficult to manage. One of the big problems is that the interviewee in many cases gives the answers that he or she thinks the interviewer wants to hear. Of course, this is well known to the professional data gatherers and very sophisticated methods are used to reduce the error due to this type of bias.

Pause for thought

Can you think of any possible uses for key performance indicators that might help you in your work? What means will you use to measure and record them? Might you use some of the statistical tools we mentioned? How accurate do they need to be? How about in your home life or sports, etc.?

Summary of learning

There might be hundreds of measures that relate to our goal, but these are not equally important. A few of them are critical and if we can identify them, these critical goals can be highlighted and separated from the rest by calling them Key Performance indicators. Surveys are yet another form of monitoring and measuring. These usually involve opinion data which is usually the most difficult to manage.

SECTION 10: CRITERIA FOR SELECTION OF DATA TYPE

When, where and how much data to collect

By now, you will probably have figured out some of the answers to these questions. In some cases, the answers are obvious and depend upon why you want to collect it in the first place. However, here are a few general rules. The answer to '**when**' is generally, as close to the time when the data is originated as practically possible. This is particularly the case with opinion data. Opinions are changing all the time and the longer the gap between whatever you are trying to measure and when it is collected will cause a possible drift in the results. The same is true for process data, dimensions can change with time, with temperature, stress relief especially with plastic materials and with possible chemical changes.

The '**where**', is similar to the forgoing and is as close as possible to the point where the data is created as possible. The larger the gap the more chance there is for contamination, manipulation, losses and other factors that might cause bias. **How** data should be collected depends on how you intend to use it. Some thought needs to be given to this because it depends upon the accuracy required, what data collection tools are available and the necessary skills in using them.

Types of data collection

There are a variety of ways in which data can be collected depending on data and its possible use. One of the most popular is the check sheet. Check sheets are simple to use, and we see them in our everyday lives.

Process measures

Process measures are those that are taken during the course of operations to detect unwanted deviations from the plan. In a manufacturing situation this might involve measurements taken by hand with gauges of various types such as clock gauges, micrometers, rulers, tape measures, air and electronic gauges, coordinate measuring machines, and both assisted (magnifiers) or unassisted visual inspection.

It also includes a wide range of non-destructive testing methods such as magnetic Ink detection, fluorescent die penetrants, echo scanning in various forms and a range of destructive tests.

It is a fact that over 80% of all inspection is by the human eye using no aids of any kind.

In many situations the measurements might also be automatic or semi-automatic. This can include direct measurements involving mechanical devices, inductance, capacitance, electrical current, voltage and resistance, high speed electronic optical systems using cameras, air gauging, etc.

In most cases other than where ratios of wavelengths of light are used or where one set of data is compared with another from the same measuring source, calibration and traceability of measurement is important especially where high precision is a requirement. In non-manufacturing, all of the same principles apply to service situations as they do to manufacture.

In non-manufacturing, all of the same principles apply as to manufacture. It is just that the tools may sometimes be different.

In fact, we can find monitoring and measuring equipment all around us in all walks of life. Probably one of the most universal these days is bar coding. This is used

Figure 3.42   Digital and bar codes

in manufacturing for defect coding, warehousing for stock control and, of course, this is also used in the retail industry extensively.

Then there are metal detectors at airports and shop exits, etc. There are all the dials and gauges that light up as soon as we switch on the ignition of our cars. There we are confronted with a range of gauges and sensor lights as is the case with aircraft, ships and boats and, more recently, GPS and satellite navigation. All of these can be regarded as forms of in process control. Many require calibration, most are subject to error both within themselves and especially at the gauge/ human interface.

Tools for data collection

Inspection and test records

The principles for the management of inspection and test records is best described in ISO 9000. Key principles are that that they should be kept securely with limited access. They must be up to date. The content must be traceable to the source of the data whether it be human, or machine generated. The collection must be authorised and carried out by a competent person if relevant. The time and date of the collection and relevant circumstances must be clearly stated. Where calibration is involved, the traceability path and the relevant test equipment, results and dates must also be clear.

Figure 3.43 shows such a display in one Japanese refrigerator manufacturing plant.

Note the bottom left green figures. This is an accumulative record of the total stoppages since the start of the shift. Behind those figures accessible on a computer monitor in the work area will be a breakdown of that 13 minutes in the form of a Pareto diagram. This will influence the choice of topics to be tackled in the quality circles meeting.

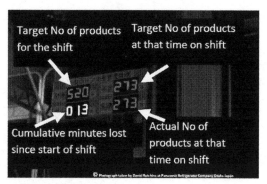

AUTONOMATION
Example of use of overhead displays

Example of ANDON type Display system I this is a powerful form of 'Visible management

Figure 3.43   ANDON display

Types of data

Visual management – storyboards

These are used a lot in quality circles and other regular forms of project-by-project improvement activities. A typical example is shown in Figure 2.31.

4 *Business performance improvement*

This chapter is designed to work differently from the others. You can, if you wish, just read it through but to get the best benefit from the content it is recommended that you do not do that. The chapter is designed to enable the reader to gain experience in the most powerful approach to problem solving you will find. Ideally, if you can get the collaboration of four or five colleagues with more or less similar work experience, work with this text to identify, analyse and solve a problem of your own choosing in your workplace. You may then present the results to either your peer group or a higher level of management. Do not select anything too complicated, but you can later when you have gained some experience. Keep it simple for now but that does not mean you cannot select a problem that is costing money, presents an ongoing hazard or is just simply a nuisance. The text will tell you what to do. The chapter is made up of the following progressive sections.

- Section 1: Introduction to problem solving
- Section 2: Defining a project
- Section 3: Project identification based on cost
- Section 4: Recording the current situation
- Section 5: Process analysis
- Section 6: The fishbone diagram and related tools
- Section 7: Testing the theories
- Section 8: Evaluating the possible remedies
- Section 9: Holding the gains
- Section 10: Presenting a project.

SECTION 1: INTRODUCTION TO PROBLEM SOLVING

What you will learn in this section:

- A brief introduction to the unit
- An explanation of what a problem is (not as obvious as most people think)
- The start of the journey from symptom to cause.

Problem identification

What is 'a problem'?

Basically, there are two kinds of problems. There are those that occur suddenly and usually unexpectedly such as a pipe rupture, measles, explosion and power cut, etc. These usually require 'firefighting' of some sort; we will refer to these as 'random' failures even if some of them occur frequently. Then there are those that are endemic, chronic or 'residual' problems that are 'built in' to the process; they are constant and are the main contributors to the underlying average deficiency in output from the target. They will have been present from the initial design of the process and result in a less than perfect output. They have possibly been present but hidden in the organisation since start-up! Because they are residual, the likelihood is that we accept them as fate and rarely give them any thought.

Therefore, we ignore them and build them into standard costing as 'waste', 'set-up times' 'absenteeism', 'sickness' and absorb them as being a part of the operation. Some refer to this waste as *muda*, a Japanese term describing seven wastes, but this is a distraction. Rather than isolating these wastes individually, it is far better to look at the process as a whole and ask, why might it not be performing at optimum capacity? If we do this, we are likely to find that these wastes do not line themselves up conveniently in little cells but rather they interact with each other making a complex maze of problems. For example, a UK aircraft manufacturer was taking an average of 72,000 man-hours to build a pair of wings for a well-known but small commercial aircraft. It was obvious that this was excessive. Using the techniques described in this chapter and using project teams, in one year it was reduced from 72,000 to 44,000 man-hours. The following year it was reduced from 44,000 to 24,000, then to 17,000. This was the result of over 300 improvement projects. If instead of following this approach, they had started from isolating the so-called wastes described in *muda*, it is highly likely that most of the problems solved in that programme would still be there if the company had survived. Because we allow for them in the standard costing process we do not notice them unless they suddenly get significantly worse. These are the special causes that we need to identify and attack. SPC will help us to do this but it is not the only way.

It is because we have no sensitivity for these endemic residual problems that the firefighting problems that are more sensational and dramatic get all the attention. It has been estimated that the average manager spends some 70% of his/her time solving sporadic problems. It makes them feel good and useful, but the chronic problems just go on and on. The sporadic ones raise the adrenaline and are memorable. For this reason, most organisations employ people who are good at tackling these problems but in consequence they ignore the others that are endemic in the process.

Notice that in Figure 4.1 the performance varies with time, which is to be expected. Most of the readings are close to the average but not those marked '1'. They are way above the average and represent the intermittent random problems we have discussed. These have been tackled by firefighting and the level has returned to

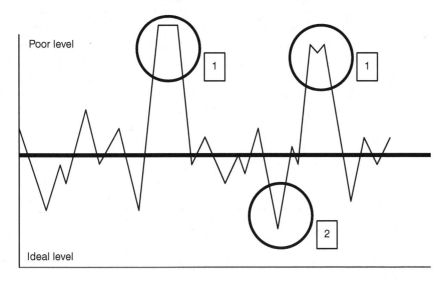

Figure 4.1   Sporadic problems

near the average. Note also the dip marked '2'. This indicates that for some inexplicable reason the process showed a dramatic improvement. The probability is that this went unnoticed but it is just as important as those marked '1'. Had we studied the events that took place at around that time we may have found the cause of this unexpected improvement and if we had implemented a change then we may have 'made happen on purpose' an event that happened initially by luck. This will help drive the average level down to become the new improved level. Rarely are these opportunities seized.

Identifying the random or chronic problems

What is most important is the average level of deficiencies. We would prefer this to be at the level of perfection, but this requires a radically different approach than that used to deal with the random problems that we have discussed. Solving the random sporadic problems is easier than is required to solve the chronic. In the case of the sporadic, we can ask 'what changed'? It was OK until an hour ago so what is different? We find the difference and if we got it right, then life returns to normal. In the case of the chronic it is different. By definition, nothing has changed. This is how it has always been, so we need to get a better understanding of the relationship between the process variables and product results. This requires an investigative study, which we will take shortly. It is a sort of detective work to find out.

In the case of chronic or endemic problems, which are keeping the average at an undesirable level, we cannot use the same 'random problem' approach because nothing has changed. This is the way it has always been and how the process was designed. The average figure represents a level of ignorance. If we knew why it was running at this level, we would have done something about it before.

Therefore, solving these problems requires a voyage of discovery. When we dig in, we find, generally, that there is not just one cause of the less than desired performance but usually a myriad of them. Many of them are quite small; a few may be large as, collectively, they will follow the Pareto principle. Approximately 80% of the poor performance is likely to be caused by just 20% of the causes. If we can isolate these and remove them, we can make a significant improvement to the performance of the process. This is what happened in the case of the aircraft wings and the improved tap-to-tap time in the steel works. Tap-to-tap time is the time it takes for a ladle of molten iron to travel from the blast furnace to the steel-making area, raise the temperature in the ladle to some 3000□ to burn out the impurities, empty the ladle and the return it to the blast furnace for the next cycle. In our case, this was reduced from an average of 220 minutes a cycle to less than 165, which was below the theoretical nominal capacity of the plant. This improvement was achieved in less than one year and was again some 300 projects, some large and many small.

It should be clear from the examples above that the improved overall level of performance is achieved largely using the project-by-project improvement process described below. However, in many cases, the methods described can also be applied to random problems as well. As these random problems are eliminated and if the fool proofing process is applied, they will also help in reducing the average as will holding the gains of the random surprise improvements. As this happens the variation around the mean is also reduced and the whole process becomes progressively more predictable. In turn, this makes scheduling easier, which will have a cumulative impact on work in progress, set-up times, waste of all descriptions and finished goods stocks. In fact, it is the structured project-by-project process that brought about the so-called lean manufacturing revolution in the first place. Sadly, this fact is often conveniently overlooked by lean protagonists.

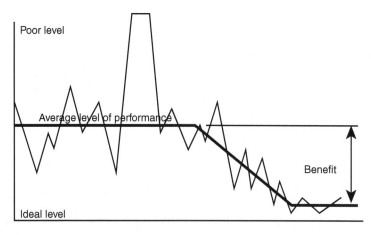

Figure 4.2   Quality improvement

Pause for thought

One good way to identify the residual problems that are inherent in the processes is to identify an output of one of the primary processes such as delivery time and note the frequency of late deliveries. Often, there will not be a single possible cause, there will be a myriad of them, and we do not know which are important. These collectively are the reason, but they will usually follow the Pareto principle. Some 20% of them will probably account for approximately 80% of the deficiency. Just for now, identify some of the output variances that impact on your sales. In a later chapter we will consider how these can be challenged.

In your private life, if you are an athlete of some sort, make a list of all the possible causes of your below target performance. Try to use the Pareto principle on those. Are there any that would appear to be the major ones that you can concentrate on to achieve better times? Usain Bolt watch your records!

We spend much of our lives solving problems. So, we probably think we are good at it.

SECTION 2: DEFINING A PROJECT

So, what is a problem in the context of continuous improvement?

To get started, it can be as simple as you like. There is no point in biting off more than you feel you can chew. This is a learning process. Once you have proved to yourself that it works then yes, eventually try to change the world! You can involve others if you wish, this would be ideal, but it is also OK to work on one on your own. The project you select should be your choice and something within both yours and the others' – if you are setting up a team – work experience. It does not matter whether or not you complete it by the time you get to the end of this chapter, but it would be rewarding for you if you do, although there is no pressure. It is entirely intended to make the learning more interesting and to learn by doing. The objective is to help make the learning as realistic as possible.

What we hope you will agree at the end of this chapter is that the methods described are not only fun to use but can result in some dramatic improvements you may not have even thought possible and with far less effort than you might have imagined. Not only that, but unlike a lot of problems that frustratingly reappear, **these problems stay solved**. Also, if you recall the 'hidden factory' idea in an earlier chapter, we will not just be solving the day-to-day problems that just pop up, but we can make some significant improvements to our processes that we might not previously have thought possible.

To get started, we need a road map for the improvement process.

The starting point on the journey is a **problem** we wish to solve. Let us suppose we have selected 'late deliveries'. The next question then is 'what causes late deliveries'? There will be multiple theories, many of which will clash with each

BUSINESS PERFORMANCE IMPROVEMENT

other. It is important to bear in mind at this stage that these are 'theoretical **causes**'. We can generally dream up many of those, hundreds sometimes. Eventually, we will need to do something to actually 'prove' whether they are really relevant or not.

When we have isolated what we think are some serious causes, what we want are some remedies but first of all we must be sure that they really are the causes and for this we must use a concept known as 'root cause analysis'. Unfortunately, this is where we nearly always go wrong, and we set ourselves a trap! The bait in the trap is the fact that we want results **now**! So, we take short cuts.

No sooner do we identify one or two 'possible' causes, to save time, we make assumptions and jump to conclusions. This is what we have always done, and it is a principal reason why the problems continue to exist. We must find ways to actually prove which are the real causes. It does not matter how obvious they may appear, experience says that you will always have surprises as will be seen in an example that follows.

Proving the theories can range from a simple matter of, say, running a process slightly faster or slower and noting the possible changes to find the optimum. At the other extreme, it could involve a complex study using design of experiments, regression analysis or other sophisticated analytical tools. Fortunately, whilst these situations exist, they are relatively rare. The majority of projects, some 90% or so, generally require nothing more than some simple tools and some common sense. Those who make a living out of selling Six Sigma and Lean Six Sigma courses will not forgive me for saying that, but it is a well-proven fact.

Remedies

Having now 'proved' the causes, the next step is to evaluate some possible solutions. Sometimes, this might be obvious. For example, if a cause is found to be that an oven is running 10 degrees hotter than is good for the process, it might simply be a case of changing the operating instructions, resetting a thermostat and briefly retraining the operator. We may, however, audit the process from time to time to be sure that things have not slipped back to where they were.

This leads to an important observation. The remedy just suggested could be 'reversible'. After the investigation it was possible for the operator to reset as it had been before. He may have had some ulterior motive for doing this, for example heating his lunch! Sounds implausible but it has happened!

An 'irreversible' solution might have been to change the controls in such a way that they are tamperproof and cannot be adjusted unless with authorised means. This is known as 'foolproofing', 'Poke Yoke' or other similar names. The important thing is that the process cannot return to its former state.

In summary

The problem-solving process is as follows: we start off with a **symptom** – the outward signs that something is wrong. Maybe your child woke up in the morning covered in spots. You would be right in thinking that you have a problem but the spots themselves are not the problem, they are a symptom. Symptom of what?

You have burned the toast (we usually call this an 'effect' but you will agree, it is also a problem. Or the car doesn't start. Something like that. (This is also a symptom, but we still tend to call it a problem.)

We can see that in every case, the words 'problem', 'symptom' and 'effect' all have the same meaning. We just vary their use according to habit. However, we will soon see that we must not confuse these with either 'cause' or 'solutions'. It sounds obvious but, in fact, it is not, we do it all the time. It is now the doctor's job to go from symptom to cause. There could be many possible causes – measles, chicken pox, too many ice creams! A skin disease of some sort.

Here he has two choices. Option 1: he might, and often does because he is busy, go direct from symptom to remedy by taking a guess at the true cause. He writes out a prescription. You take the medicine and if you are lucky the spots clear up but, on many occasions, maybe you are not so lucky.

Instead of getting better the spots get worse and other symptoms appear. The doctor might do the same thing again and keep on doing it until if he is lucky he hits on something that works. The spots clear up and life returns to normal. However, the child could also have died in the process! Initially, it might have seemed quicker for him to guess the cause but having got it wrong, maybe it would have saved time in the long run to have got you properly checked out in the first place. Do not criticise the doctor, all of us are doing this all the time.

Here is an ALTERNATIVE approach. The doctor decides on option 2. Instead of just writing out a prescription, the doctor identifies as many theories of causes as he can from experience.

It could be: an allergy (a lot of possibilities here), chicken pox, measles, small-pox, a skin infection, etc. He now has to test these theories. In the case of the allergy theory, has anything unusual happened recently? Yes, we have a new kitten. Hmmm, maybe the child is allergic to cats. It is quite common. Anything else? Well, oil seed rape is in flower right now. It could be that but it is more likely to result in a throat infection.

How about chicken pox? Do the spots look like chicken pox? How did it start? Does the child have a temperature? First, the doctor is trying to put the theories in order of maximum likelihood and later, he/she may decide to test the main theories by giving a blood test or taking a swab, an X-ray or CT scan, etc. Sooner or later, in most cases, he/she does find the true **cause**. The diagnostic stage is now over and we move to the next stage.

Cause to remedy

The journey from cause to remedy is usually simpler than the journey from symptom to cause. The reason is that once the cause is known, the skills required are much more obvious.

To be a 'problem' the test is that it must be both **undesirable or nasty _and_** you must be able to count the number of times it happens or measure it in some way.

Think about it. **If you cannot count it or observe it happening, you will have no evidence to show that you have solved it** or by how much things may have been improved.

How will you prove it to anybody and who will believe you?

Pause for thought

Think of some situations in your own life, work or home, where initially you have jumped to conclusions and got it wrong and wasted a lot of time before doing a proper check or maybe just given up and put up with the problem!

So, if you plan to run the improvement process alongside this chapter. Check the following.

The answer to the question in Figure 4.3 is none of them. They only seem like problems. If you were challenged and told 'poor training' that is not a problem. You might respond, yes, it is, we do not do anything like as much training as we should. Other people do far more than we do. Maybe they do, but that is not important. Now, for the next question. OK, what is it in the company that is not good because of insufficient training? You might respond, high labour turnover, excess scrap or rework, computer crashes, customer complaints, etc. Good. These are all 'problems'. They are countable or measurable and nasty. Problems. Just take one of them. Excess scrap. What are the possible causes? Poor instructions, no air con, too noisy, lack of training, etc. Got it! So, lack of training is a possible cause, not a problem. We can deal with that. First, we must prove it. If it is correct, then the next question is what is possibly wrong with the training? We check that, and it proves correct. So, we can now think about a solution. What training might be necessary, how long should it take, what might be the content, who should be trained, etc.? If we get all of that right and implement it, then if we were correct then we would expect scrap to reduce. If not then we missed the true cause, lack of training was a red herring and off we go again. If we had not done that, maybe if our argument had been persuasive and the organisation had some cash to spare and allowed an increase in the training budget, then the next question would have been, training in what and for whom? We should spend it or it might be withdrawn next year.

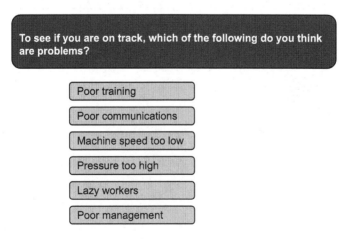

To see if you are on track, which of the following do you think are problems?

- Poor training
- Poor communications
- Machine speed too low
- Pressure too high
- Lazy workers
- Poor management

Figure 4.3   Identifying problems – 1

Pause for thought

Is there a ring of truth in this argument? It is just as true for all the other 'causes' confused as being 'problems'. By not being able to link them specifically to real problems, we can estimate the cost of our perceived remedy, but we cannot articulate its merits because there is no causal link.

Remember – you cannot solve a cause! This is a very common mistake that people make, and which gets them confused when attempting root cause analysis.

Sometimes, you might be able to prefix the 'what we perceive to be a problem' with the words 'we need', for example.

If you can prefix the perceived 'problem' with the words 'we need a' then straight away, you will know that you are not dealing with a 'problem' but with a 'solution'. A solution to what? You can ask. What software are you talking about? What do you think will be improved if we invest? Maybe the answer is 'well, we are being criticised for late delivery of the monthly profit and loss figures'. If we had new software, we could do it more quickly. OK, we might then say, so the problem is 'late delivery of P&L accounts'? Yes, OK, let's test this and see if you have correctly located the cause. There is an example later in this section and surprise, surprise, the cause had nothing to do with the software or the computers used. If an investment had been made, the problem would have remained right on dead centre.

Now, Figure 4.5 shows some real problems. Why? Because they are all countable and all undesirable. We can count the number of times they occur. Later, their absence will show convincingly that our solution worked.

Pause for thought

How many 'problems' can you think of that everyone talks about that are not problems but either causes or solutions? Do not forget, you cannot solve a cause and you cannot solve a solution!

To solve problems, we need a problem-solving road map.

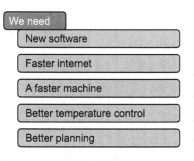

We need
New software
Faster internet
A faster machine
Better temperature control
Better planning

If you can do that then again, these are not 'problems' either, they are possible solutions but solutions to what?

They are, however, pointers to what are the real problems as we will see.

Figure 4.4   Identifying problems – 2

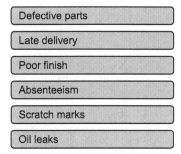

Figure 4.5   Identifying problems – 3

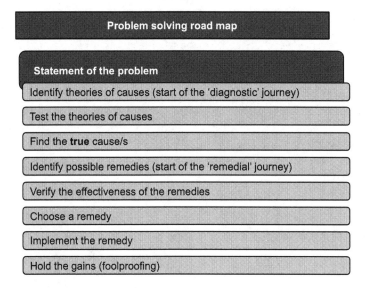

Figure 4.6   Problem-solving road map

Pause for thought

Consider an area for improvement that you are familiar with, and write a statement of the problem. Also include how much you think it costs or what is the undesirable outcome. Then a goal of what you hope to achieve and what success looks like.

Description of the problem

Remember, earlier we said there are two alternative terms used for problems and they vary from one situation to another, but they are all problems and can all be treated in the same way. We saw that in the medical profession they might refer to something as a '**symptom**' – spots, cough, etc.

We also saw that, on the other hand, if burnt toast popped out of your toaster, it is still a problem but you might refer to it as an '**effect**', for example, the effect of having the thermostat turned up too high. These words are interchangeable, but it is best to fix on just one of them in the project.

Have a go and work through a project right now!

The journey from symptom to cause/s

This is a twostep process.

- Step 1. Identify theories of causes
- Step 2. Test the theories to find the true cause.

Step 1 typically (but not always) involves the use of techniques such as process analysis and the fishbone diagram, both of which are explained in detail later.

The journey from cause to remedy

This is a three-step process.

- Step 1. Identify possible solutions
- Step 2. Test the solutions
- Step 3. Select the best solution bearing in mind cost/benefit, reversible/irreversible (also covered in our case study).

Project completion

This involves finding the best means of holding the gains by fool proofing. If a reversible solution has been found, then it will be automatically fool proofed. After which, we present the results

SECTION 3: PROJECT IDENTIFICATION BASED ON COST

What you will learn in this section:

- How problems can be identified (more on this in Section 5)
- Brainstorming
- Prioritisation matrix
- More on quality-related costs.

Problems can be identified several ways.

1. Ask people.
2. You may be assigned a problem as a project. If this is the case you may need to follow the advice above and convert what might be stated as either a **cause** or a **solution**, into a **problem**.
3. Conduct a brainstorming session

Brainstorming or thought showering?

We apologise to those who do not like the former term, but it has been used so extensively across the entire spectrum of management books for so many decades that, like it or not, it will not go away. It is a technique that you can find useful in a wide range of situations, but you are free to call it what you like.

Round robin method

Rules Set a time limit (typically 20 minutes). If this is not long enough, then have a break followed by another session.

- We want many ideas
- Write down all ideas as they are suggested
- Take turns
- One Idea/person/turn
- Say 'pass' if you need time to think
- No criticism during brainstorming
- No censorship during brainstorming
- No discussion during brainstorming.

Use TPN to make a project selection.

Review each brainstormed idea to:

Clarify understanding where necessary, and re-title the bullet point where appropriate.

Assess the extent to which these problems (opportunities for improvement) are group controllable.

Code:

- Totally group controllable (T)
- Partially group controllable (P)
- Non-group controllable (N).

Use TPN to make a project selection. Sometimes the 'P' can be turned into 'T's if the appropriate resources can be added to the team.

We can number our list of 'T's as follows:

1. Broken needles
2. Oil stains
3. Chain marks
4. Vibration
5. Broken yarn
6. Out of adjustment
7. Lack of tension
8. Water stains
9. Stock variation
10. Worn shaft
11. Overheating
12. Wrong instructions.

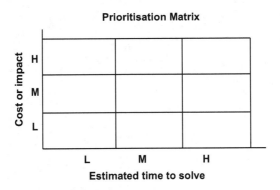

Figure 4.7   Prioritisation matrix

From our selection of totally solvable problems we can use the prioritisation matrix to make a choice of project. But there are other methods such as paired ranking.

The prioritisation matrix is a subjective technique but is very popular in problem solving.

Project selection

Rate the relative strength of each potential project with respect to: measure of potential impact. What is the project's potential impact? Usually, measures should indicate the project's impact to:

* Retain customers and attract new ones
* Reduce the costs of poor quality

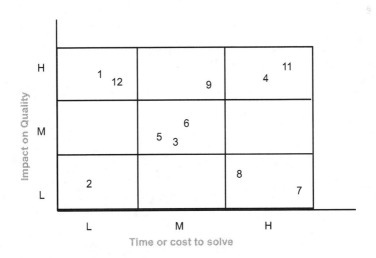

Figure 4.8   Prioritisation matrix – complete

Evaluate each possible 'solvable' project to validate that is:	
Chronic rather than sporadic	The project should correct a continuing problem, not a recent specific episode.
It should be high on the list shown by Pareto analysis (we are about to show you this technique)	
Significant	When a project is completed, results should indicate the effort was worth the effort.
Of manageable size	Most projects should take no more than six months to a year to complete. In the case of this chapter estimate two to three months. Projects that would probably take longer should be divided into smaller projects likely to yield results more quickly.
Likely to make an impact	If the organisation is new to quality improvement, it is important to look for a project certain to produce a successful outcome.
Able to be measured	All quality improvement projects require measurable problems. If the organisation is new to quality improvement and does not have data to evaluate the potential impact of a first project, measurements should be taken by the project team before it proceeds.

Figure 4.9   Project evaluation

- Provide return on investment
- Enhance customer satisfaction
- Enhance employee satisfaction.

How urgent is the project to the organisation? Typically, urgent projects address quality problems in core services, problems that make the organisation highly vulnerable to the competition or issues that are crucial to key customers.

SECTION 4: RECORDING THE CURRENT SITUATION

What you will learn in this section:

- Process mapping (recording the current situation prior to carrying out process analysis
- Introduction to process analysis.

Process mapping

This and the subsequent technique, process analysis, are probably amongst the most powerful of all the investigative techniques in the problem-solving process. Please take time to study it.

Let us imagine that we have selected the problem of **'late delivery'**. It is actually a very popular topic! In this case it was not 'late delivery' of a product to a customer of the company, it was late delivery by the accounts department to the board of directors. They are an 'internal customer'. The accounts department in this case are the 'supplier'.

Late delivery meant considerable criticism of the finance department from the directors. The finance department thought this was unfair because they did not think it was their fault. They did everything they could to ensure prompt delivery. No one else in the process thought it was their problem either. They all thought the problem was caused by the computer process. Nobody had ever proved this but, equally importantly, nobody had disproved it either. It had been repeated so many times over the years that everyone simply believed it. We decided to test it. This problem was a major concern of the finance department.

We used the technique called process mapping and process analysis, which we developed in DHI in the late 1970s.

It was found that the variation of time was greatest at four stages. Particularly, when the data sheets were in the operations departments, and, second, at the point where the data was keyed in to the computer. When these were investigated further it was found that the reason for the long variation in the departments was due to the managers either not knowing how to fill in the forms correctly due to bad form design or that they did not want to do it at all as, according to them, the job was low priority compared with other pressures on their time.

There was also a problem of poor handwriting, use of wrong terminology to describe activities and so on.

There were two possible solutions.

1. To send everyone on a handwriting book and to be taught what is the correct terminology for the various activities.
2. Redesign the forms. This could involve the use of tick boxes to reduce or eliminate the use of handwriting, force use of the correct terminology and to make the forms quicker and easier to complete.

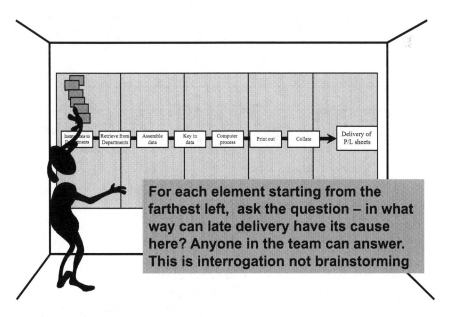

Figure 4.10   Process mapping

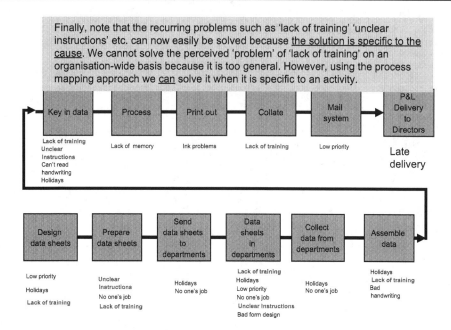

Figure 4.11   Analysing theories of causes – 1

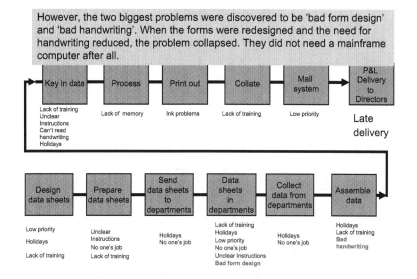

Figure 4.12   Analysing theories of causes – 2

The first of these two remedies were considered 'reversible' and people could easily slip back into their old ways. Maintenance of this solution would require constant auditing. The second remedy was 'irreversible'. Once the old forms had been destroyed there was no way back. This is known as 'fool proofing' or in Japanese *poka-yoke*. Naturally, the second option was the more popular.

Pause for thought

Consider a simple process that you are familiar with and generate a process map. Consider how efficiencies may be saved if the process was improved.

Summary of learning

What you have learned in this section:

- How process mapping can be used as an effective way of recording the situation that existed before a project is started so that reference could always be made to the circumstances before improvement as a way of comparison and also to help define the scope of a project.
- A brief introduction to process analysis before it is covered in the next section and to show how recording the time variation of certain key activities can lead to the identification of key causes.
- How process analysis is probably the most powerful tool available to identify the possible causes of a problem at any stage in a process so that nothing is overlooked. Also, to demonstrate that despite long-held beliefs, the true causes of a problem can turn out to be very different from what had been always assumed.

SECTION 5: THE FISHBONE DIAGRAM AND RELATED TOOLS

What you will learn in this section:

- The use of the fishbone diagram
- Pareto analysis
- The construction of Pareto diagrams.

We commence the section with a review of the technique called the fishbone diagram, also known after its inventor as the Ishikawa diagram. It is often also erroneously called the cause and effect diagram. Of course, that is what it is, but this term can be confusing as process mapping is also a form of cause and effect analysis. When diagnosing the cause of a problem, a cause and effect diagram whether it is process analysis or the fishbone diagram, helps to organise existing theories about causes and develop new ones. A cause–effect diagram cannot identify a root cause; it simply presents graphically the many possible causes that might contribute to the observed effect. This technique helps focus the search for the root cause and contributes to the team's understanding of the problem.

A fishbone diagram helps a team organise theories for systematic review. In addition, the diagram often challenges team members to come up with new theories by asking 'Why?' for each factor they list. For example, in the sample diagram (Figure 4.13??), the team might ask why there could be dirt in the tap. The team thought that the possibilities were jointing hemp and wire wool not properly cleaned out by the plumber or debris not cleaned out during the tap manufacturing process.

Pareto analysis

In order to prioritise projects, we can make use of the tool provided for us by Dr Juran, which is known as Pareto analysis. He gave it that name in recognition of Dr Wilfredo Pareto, an Italian American economist who discovered the 80/20 principle. It was Dr Juran who first saw its connection with quality-related problems in the 1950s and it has retained its label ever since.

Wilfredo Pareto related this distribution to wealth and found that in most countries at that time approximately 80% of the wealth was in the hands of some 20% of the population. Dr Juran found that not only did this principle apply to wealth but, indeed, to a very wide range of topics including, generally, the cost of poor quality. Figure 4.14 shows a typical Pareto diagram. Notice how attention is drawn to the largest column on the left. This figure shows far more dramatically than a list of numbers that the biggest problem or the most frequently occurring event is item D. The upturn in the last column occurs when a 'miscellaneous' column is included for the small items.

For example, if you have made a longish list of quality related problems, or, better still, brainstormed the list with some colleagues and then estimated the cost of each, you will undoubtedly discover that the Pareto distribution will apply to them. There will be a few that cost a large amount, this would be the 20% that accounted

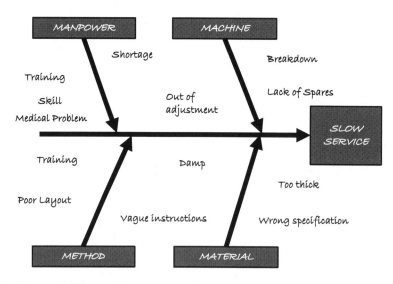

Figure 4.13   Fishbone diagram

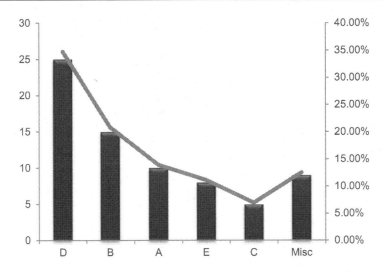

Figure 4.14   Typical Pareto chart

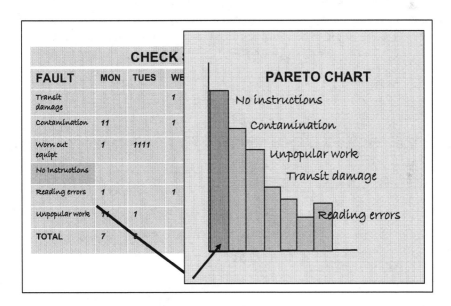

Figure 4.15   Check sheet to Pareto chart – 1

for some 80% of the cost, then there would be some 80% that collectively accounted for only 20% of the cost

Merits of Pareto analysis

- For those concerned with performance improvement – highlights the key problems

- Useful way of communicating results to team members
- It makes a big impact in a management presentation
- Problem solvers can compare similar data obtained after implementing
- Goal-setting mechanism, which concentrates on important problems
- Those concerned with performance improvement will be impressed
- Problems on the left tend to be management solvable
- Problems on the right tend to be worker solvable.

Pause for thought

What is the purpose of the fishbone diagram? Might you use it in your work? If you are working on a project, then practise the technique at the appropriate stage and use the rules described in brainstorming.

Summary of learning

What you learned in this section:

- The section demonstrated how the use of the fishbone diagram is an effective way to enable ideas generated in a brainstorming session to be grouped easily as the session runs to enable clear analysis afterwards.
- How Pareto analysis can be used to highlight the main possible causes for further analysis and how Pareto diagrams are constructed.

SECTION 6: KEY TOOLS FOR TESTING THE THEORIES

What we will learn in this section:

- The use of check sheets
- How a check sheet can lead into the use of Pareto analysis
- How a check sheet can be used to verify the importance of possible causes identified on a fishbone diagram
- Risks in taking samples
- Use of charts to verify theories of causes.

Earlier, we looked at a range of tools available for collecting data in order to test theories of causes. Here, we will remind ourselves of the most commonly used ones for testing theories of causes but there are a great many more and these can get quite sophisticated in rare cases. However, there is no need to panic if this happens in some projects as there is a way to deal with it.

A check sheet is one of the most common data collection tools used by project teams to verify the frequency of either symptoms or possible causes. It is simple to use but care must be taken to ensure that sufficient data is obtained for the degree of certainty required.

CHECK SHEET

Description of cause	Mon	Tue	Wed	Thurs	Fri	Total	Cost
A							
B							
C							
D							
E							
Misc							

The results of a check sheet can be used to create a Pareto Diagram.

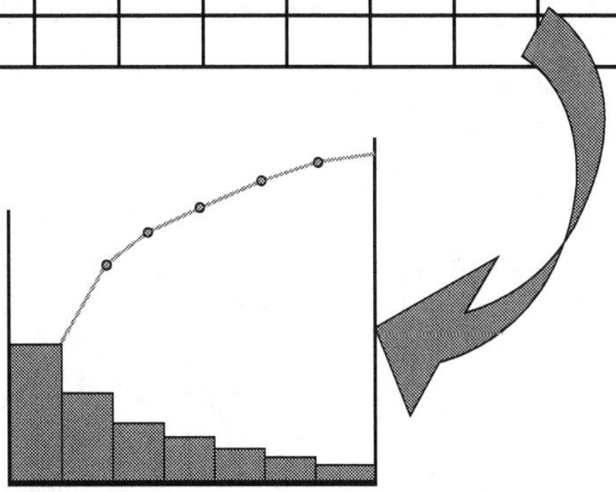

Figure 4.16   Check sheet to Pareto chart – 2

The team should always be open to new theories even after using process analysis or the fishbone diagram. Often, the results of data analysis will suggest new theories that should be added to a cause–effect diagram or process analysis. The team must decide how to test them further.

Data collection plan

Which theories are supported by the results? The team should highlight these on its cause-effect diagram or process analysis and plan for any further testing that may be needed to prove the root cause. (Later in this programme we will discuss in more detail how to recognise when the root cause has been proven. Which theories are eliminated by the results? The team should remove these theories form further consideration by crossing them off its working copy of the cause–effect diagram or process analysis sheets. Root causes should account for most of the problem. Any theories that account for only small parts of the problem should be eliminated.

- Remember the rules for data collection, for example
- The risk of saying something is OK when it is not
- The risk of saying something is wrong when it is OK
- Risk is determined by sample size
- Reduce bias due to sampling error, human error, subjective data, etc.

The basic steps are as follows:

- Design the data collection
- Train data collectors

- Conducted some tests to establish the main cause/s (do not forget to ask those who are in the process as they will also have useful information).

Conduct root cause analysis on the most likely causes identified

There are two questions that will help you decide whether you have found the root cause.

Does the data suggest any other possible causes?

After each data collection and analysis, it is usually possible to discard some theories and place more confidence in other. Theorising is not a one-time activity, however. Each data display – the Pareto diagram, histogram, scatter diagram, or other chart – should always be examined by asking whether it suggests additional theories. If you have competing plausible theories that are consistent with the new data and cannot be discarded based on other data, then you have not arrived at the root cause.

What new theories are suggested by the results? The team should always be open to new theories.

Do not forget, your team are not the only people who know things. Even in processes that you think you know very well, there may be things that others know that you may not. Of most importance ask the people who work in the process. Remember the saying 'the guy that knows that 25 sq feet of floorspace best is the guy who lives there'!

Is the proposed root cause controllable in some way?

Some causes are beyond our ability to control, like the weather. We can control the effects of the weather by turning up the heat or running a humidifier, but we cannot control the weather directly. So no useful purpose is served by testing theories about why the weather is cold. Other possible causes are too broad and general to control and need to be broken into components. For example, 'lack of training' as a cause. We have mentioned this one several times but only because it is important.

What we have learned in this section:

- This section introduced the check sheet technique, which is amongst the most popular tools for the collection and analysis of data when testing the theories of causes.
- We saw how a check sheet can be used to construct a Pareto chart and how it can be used to verify the importance of possible causes.
- We also saw why it is important to be aware that all sampling carries risks such as bias and also the risk of saying that something is one thing when it is something else and to reduce this risk, we have to make sure we collect enough data. We also saw how the use of charts can be made to verify theories of causes.

SECTION 7: EVALUATING THE POSSIBLE REMEDIES

What you will learn in this section:

- Evaluating possible solutions based on cost or other criteria
- Fool proofing (or mistake proofing if you prefer it)
- Identifying reversible and irreversible solutions.

Evaluating possible remedies

Total cost

Costs to implement a remedy should not exceed available resources. Usually, quality improvement lowers costs, but some initial investment may be necessary. Resources for that initial investment may be acquired by:

- Reallocating existing budget and staff
- Obtaining new budget authority from the quality council
- Drawing on existing funds available to quality improvement teams for implementing remedies.

Impact on the problem

The team needs to estimate the impact of alternative remedies on the problem. Some remedies may solve more of the problem than others.

Benefit/cost relationship

While total cost and impact are important considerations, the cost of each alternative compared to its impact on the team mission is even more important. A remedy with an unfavourable benefit/cost ratio is a poor solution.

Pause for thought

- Consider an area for improvement
- Review all the tools in this section for the appropriateness for use
- Create a table indicating the positive and negatives of use of that tool in your situation.

Reversible or irreversible situations

When we investigate this, we discover that solutions may be of two kinds.

- Those that are reversible
- Those that are irreversible.

Where possible, it is best to select from those possible solutions that are **irreversible**. A typical example of a 'reversible' remedy is one where we issue instructions or give training to a worker to perform the task in a different way. Initially, they may follow the advice but all too often they may revert to old habits. If a reversible solution is the only option or is considerably less expensive than an alternative, then the solution would be to include a check on it in the internal auditing process.

An example of an irreversible remedy might be: instead of telling an operator to run the machine at a different speed, to perhaps change the gearing or something like that so that it can only run at the required speed.

Culture/impact resistance to change

When evaluating alternative remedies, the team must consider the impact of each proposed remedy on those who would be affected by it. Potential resistance to change is not sufficient reason to rule out a remedy, but it should be weighed with other factors. All other factors being equal, the remedy likely to provoke the least resistance is preferable. It makes sense to attempt to find out the reasons for the resistance and deal with them if the remedy poses no threat to any one

Implementation time

The team will want to assess the time it will take to implement the remedy and weigh this against the urgency to reach a solution. The greater the urgency, the more important the time element.

Uncertainty about effectiveness

Even if a proposed remedy has a favourable benefit/cost ratio, it may not be a good solution. For example, a remedy may require untested technology or major organisational changes. Even if the costs are relatively low and the potential payoff is high, the uncertainty of the payoff may be too great.

Health, safety and the environment

No proposed remedy should pose new threats to the health and safety of customers, the community, or people working within the organisation. The environmental impact of a proposed remedy should at least be neutral and, if possible, positive.

Summary of learning

- In this section you learned how to evaluate how to compare alternative remedies based on the use of a number of criteria
- How fool proofing (or mistake proofing if you prefer it) can be an important factor in the selection of remedies or solutions.

SECTION 8: IMPLEMENTING THE POSSIBLE REMEDIES

What you will learn in this section:

- How to implement solutions that stand the test of time!
- The use or remedy analysis to select the best solutions.

Well, having been on the planet for so long, we think the little hedgehog has made a good attempt at solving the problem of too many predators. The prickly coat might not solve all of his problems, but they must go a long way towards it so long as he stays off busy roads. When we evaluate solutions to *our* problems it might be as well to think about the hedgehog. Until the automobile came along his prickly coat solution had served his purpose since the time of the dinosaurs, so he has done fairly well. Maybe now, though, a rock-hard shell might be better!

After evaluating alternative remedies, the team agrees on the one that is most promising. Sometimes, a team might combine certain features of several proposed alternatives, drawing on the strengths of each. The matrix on the following slide can be used for evaluating alternatives. You can use it to assess each proposed remedy according to the evaluation criteria. Place an H (high desirability), M (medium desirability), or L (low desirability) to indicate the relative degree of expected impact.

SECTION 9: HOLDING THE GAINS

The final step of the quality improvement process is holding the gains. This step encompasses three activities.

Process Improvement Solution Analysis							Form 3.2
Cause	Possible Remedy	Reversible/ Irreversible	Negative Reactions	Counter-measures	Cost of Implementation	Cost Benefit	Recommended Y/N

Figure 4.17 Solution analysis

- Design effectiveness quality controls
- 'Fool proof' the remedy
- Audit.

After all the effort to improve quality, a project team needs to be sure that the gains continue, and the problem does not recur. As a rule, the team will not continue their meetings to ensure the gains are held. The team can help see to it that improvement is maintained, by implementing the three activities for holding the gains.

Pause for thought

- Why is it important to evaluate a project?
- How would you encourage others to sustain improvements?

Summary of learning

What you have learned in this section:

- How, having solved a problem it stays solved
- How to select good solutions using remedy analysis and to gain maximum support for the result.

SECTION 10: PRESENTING A PROJECT

What you will learn in this section:

- Why presentations are important
- Who is involved in presentations
- How to plan and prepare presentations
- How to conduct presentations
- Overcoming the problems in presenting projects
- What happens afterwards.

The final stage

So, the project is completed. All that is left now is to sell your ideas to others. You have done well, you are pleased with yourselves, what could possibly go wrong now? What indeed. Well actually, this is the most critical time in the process. You are possibly faced with the biggest challenge of all. We call it 'resistance to change'. There can be many reasons for this and it can come from different sources. It can even come from the very people the change it is intended to help. So, we need to give it some thought – a lot of thought because you are now in the process of selling your ideas to others.

Selling the improvements

You have a variety of ways you can do that.

1. You could write a report and circulate it.
2. You could just quietly go around and on a one-to-one basis share your ideas with those affected and leave it at that. The team in our case study did that when they shared the results of their study with the managers. The managers then cooperated in helping design the remedy.
3. The team could attempt to arrange a grand presentation of their ideas.
4. What are the relative merits and issues?

1. Writing a report

This is probably the least effective way you could communicate your improvements and it has a very high likelihood of failure.

- You cannot be sure that anyone will read it.
- If they do read it, they may not fully understand it or properly appreciate the benefits that might be achieved
- They will not get a good idea of the amount of work that went into it, or the teamwork involved.
- It is only one-way communication and unless a meeting is arranged, there is a high likelihood that it will keep going to the bottom of the pile of things to do.
- This can be very demoralising and will kill the idea of anyone carrying out future projects.

2. Meeting those impacted one to one

This is probably marginally better than simply writing a report assuming that you can get the time of the would-be participants and a suitable venue for each to meet.

3. Formal presentation

This is by far the best option for several reasons.

- It gives the team the opportunity to put on a show. In terms of giving recognition, this is ideal.
- It gives the team the opportunity to prepare the presentation, create impressive visual aids to ensure the audience really understands the project, can see the steps that were taken, the tools and techniques used
- The audience can include everyone directly impacted by the project.
- If all participants are encouraged to participate it demonstrates the power of teamwork.
- Participating in a presentation is a people building process as the participants visibly grow in confidence.
- A positive result will encourage them to continue their development and encourage others to 'get in on the act'.

The final stage

Following the presentation, it would be a good idea to practise some of the ideas that you learned in Chapter 2 on the topic of 'winning teams'. One of the most valuable and important aspects of an improvement projects is to evaluate the success of the project.

- What worked well?
- What could have been done better?
- What tools and techniques worked well for the organisation?
- Did the project have full support?
- What recognition is required?
- Are remedies sustainable?

Consider questions such as the above to determine success of the improvement project. For this reason, it is a good idea always to have a post presentation 'wash up meeting'. Closing out a project when everyone is on a high is important.

Planning the presentation

Remember, your presentation is the point where you sell your achievements to others. It is the point where you hope to get recognition for all your efforts and to obtain the support of others outside your team. It would be very unfortunate if, therefore, this did not go well.

Summary of learning

What you have learned in this section:

- Why presentations are more successful than passive report sending
- Why it is important for all team members to participate and why a wide audience is a good idea
- To be able to determine the target audience and how to plan the event with them in mind
- The best way to present a project with everyone on the team having a role
- Overcoming problems of fear of making public presentations and what happens after the project is completed including a wash-up session.

---

5 *Quality management systems*

SECTION 1: INTRODUCTION TO QUALITY MANAGEMENT SYSTEMS

What you will learn in this section:

- Difference between formal or informal management systems
- The principles of *Hoshin Kanri*
- The role of PDCA.

All organisations whether they like it or not have a management system. This can either be formal and based on some pre-existing models or it can be totally informal and based on the logic of the owners and directors. But we could summarise all of this and say, 'it is the way we do things around here'. Some organisations will conduct serious research into the choice of management system. They may elect to take a standard one such as ISO 9000 off the shelf, or possibly rely on the content of one of the National or International Quality Award criteria. Or, they may decide to create one of their own. We would not be so presumptuous as to comment on which is best for any individual organisation as all organisations are responsible for making their own choices.

However, our experience indicates that all the successful organisations in the world with long-term sustained track records design their own management systems to suit their specific needs, albeit with more than half an eye on what everyone else is doing. Almost all these custom-made approaches can collectively be claimed to adopt what have become known as *Hoshin Kanri*. It is, as you would imagine, a Japanese name. So, we have dedicated this chapter to explaining in broad terms the basis for these personalised programmes rather than repeat the detailed requirements of the Standard Versions which, if they are used by your organisation, can be explained by those responsible. It is a fact that if an organisation follows the advice we give here, then it will almost automatically result in the successful application of the requirements of the others anyway plus a lot more.

Use of the PDCA concept

You will by now be well acquainted with the PDCA concept having been introduced to it in earlier chapters. This is unavoidable because it is such a versatile concept and is hidden in various guises everywhere. This concept is a vitally important aspect of successful management systems. In fact, we would go so far as to say that if the system is not based on sound PDCA practice, it will fail to deliver the desired results. When it is present, the model will look like that shown in Figure 1.27 Nested PDCA Loops.

When properly organised, it should be possible to trace the plan, all the way down through the organisation, layer by layer in an unbroken chain. Any breaks or weak points in the chain will threaten the entire plan. However, as the plan travels down, the detail is reinterpreted by the local managers to suit the dialect of their respective departments. For example. At the top level, the plan might have an element that says 'improve yields'. That will mean nothing to a fork lift truck driver in the warehouse but when it gets down to him it will be in the language of 'broken pallets', 'oil stains on the floor', 'flat battery' and picking errors, etc.

So, through this section we will be looking at the design of management systems from first principles. At the end of this section we will briefly explain the rudiments of each of the most common 'off the shelf' approaches so that if they are being used in your organisation you will hopefully have a better idea as to their objectives.

Because *Hoshin Kanri* is used by virtually all the world's leading competing organisations, and as we have emphasised throughout this book, it involves the participation of all personnel in an organisation **and** its supply chain, we have focused mainly on that. However, it is worth noting that if *Hoshin* is carefully implemented it will contain all the concepts included in each of the others whether they are awards or standards.

As you work through this section, you will find that in many ways, it brings together much of what you have learned in the earlier chapters as these are all part of a powerful interactive whole.

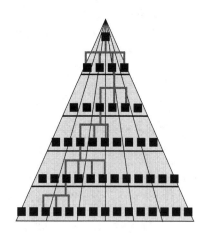

Figure 5.1   Chain of command

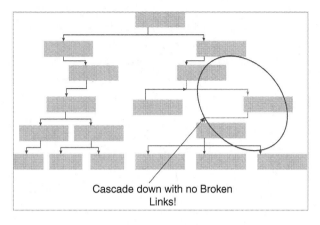

Figure 5.2   **Broken links**

Pause for thought

Check which approach your organisation is using and see if you can find out how it compares with what you are learning here.

When you have finished this chapter, it should have enabled you to see the relevance of all the content of the first five chapters if you want your organisation to flourish and for your job to be regarded as meaningful not only to others but principally to you. Your work is a significant part of your life. It is what you do, and it is what you are known for. Your job whatever it is, is just as important as any other whether you are the managing director or anyone else.

Summary of learning

What you have learned in this section:

- Why, whether the organisation has given it any thought or not, it will nevertheless have a quality management system as it is just the 'way we do things around here'.
- However, the majority of organisations will either have opted to take one off the shelf such as ISO 9001, based on one of the awards such as the Baldrige award or EFQM award or design their own such as would be required by the Japanese Deming award criteria or *Hoshin Kanri*. In this section you learned the sort of features that will distinguish between them.
- You learned that the best of the self-designed systems are generically labelled *Hoshin Kanri* but not all of them use this title. You can create your own as people such as Toyota have with the 'Toyota Way'. You

do not have to copy anyone else, but it makes sense to check out what they are doing and what you might learn from it.
- You also had the concept of PDCA reinforced as being fundamental to all good management systems.

SECTION 2: INTRODUCTION TO *HOSHIN KANRI*

What you will learn in this section:
- The philosophy of *Hoshin Kanri*
- Eliminating a 'blame culture'
- Establishing the roles at different Levels of organisation
- Top-down implementation or bottom-up?

In order to describe the concept of *Hoshin Kanri* graphically, it is common to use the analogy of a magnetic field. Here as most of us know, if iron filings are placed on a piece of paper and a magnet placed underneath, then the filings will assemble themselves as shown in Figure 5.4, in alignment all pointing towards one or other of the two poles. This is exactly the objective of *Hoshin Kanri*, only, in this case, the employees at all levels in an organisation represent the iron filings. Unfortunately, as shown in Figure 2.1, invariably, our organisations are not in such alignment. On the contrary, in many cases they are in conflict with each other. *Hoshin Kanri* is intended to change all of this as is also shown.

The diagram on the left in Figure 2.1 (originated by the author in 1976!) is a typical example of many organisations in the real world today. In fact, competition between departments is often great than competition in the market place. In the worst cases, there is a strong blame culture which was discussed in Chapter 2. The challenge of *Hoshin Kanri* is to create a culture where the arrows realign eventually to the situation of those on the right where everyone is working towards making their organisation the best in its business.

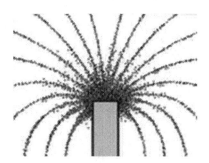

Figure 5.3   Magnetic field

Surprisingly, getting from the left to the right diagram (Figure 2.1) is a lot easier than many might think but, it must be done properly, and it takes a lot of time, several years in fact. It is not like a flick of a switch and hey presto – change. However, do not be disheartened, significant benefits can be achieved very quickly, even within weeks of starting. Also, the most important thing is where you are between those two extremes and your competitors. If they are ahead of you, you have a problem but also, it is the rate that you and they are moving to the right that is important too. If you are moving faster than them and they are not watching you, you are in luck! The hazard is that the process is quite vulnerable for a long time until it just becomes 'the way we do things around here'! Until then, any market shock or other upheaval provides a threat. Be patient. Figure 5.4 shows what the implementation process looks like when it is complete.

When complete, the development of a *Hoshin* Based Management System looks like Figure 5.4. Note that at the top, we have 'vision', goals and targets. In the middle, much the same thing but it is scaled to departmental level and at the level of direct employees, predominantly quality circles.

When implementing *Hoshin Kanri*, it is best to follow the same sequence. So, we begin by agreeing the corporate vision. Japanese experts use an analogy to explain why they think this is the best approach. They say that if you intend to sweep the stairs, it is best if you start at the top. If you want to light a fire it is best to start at the bottom. **They say that implementing *Hoshin Kanri* is more like sweeping the stairs!**

The reason they say it is more like sweeping the stairs is that you are less likely to meet with the resistance of any of the managers. The secret is that you do not take it down to the next layer until you are sure that it is being properly supported by the layer above. If you start from the bottom and try to work up, you will often reach

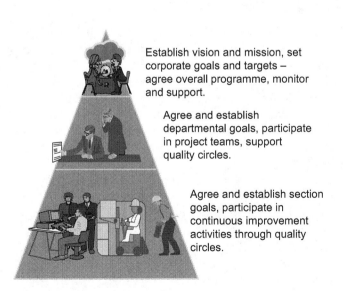

Establish vision and mission, set corporate goals and targets – agree overall programme, monitor and support.

Agree and establish departmental goals, participate in project teams, support quality circles.

Agree and establish section goals, participate in continuous improvement activities through quality circles.

Figure 5.4   Policy deployment

a manager who resents not having been involved before and who then puts up a lot of resistance.

Pause for thought

Not only in your work but also in your home life, can you see situations where a clear, measurable vision would be useful as a way of targeting efforts?

Summary of learning

What you learned in this section:

- You learned the essence of good *Hoshin Kanri* and how it is deployed through the organisation
- The importance of using the *Hoshin* approach to eliminate a 'blame culture'
- How to establish the roles at different Levels of organisation
- The relevant issues relating to top-down implementation or bottom-up.

SECTION 3: CREATING THE VISION

What you will learn in this section:

- Who creates the vision and mission statements?
- How vision statements are created
- The identification of the drivers.

Probably most people studying this book, will have been exposed to the concept of 'vision' at some time. Unfortunately, the concept is not well understood with the consequence that a large number of so-called 'visions' are nothing more than wishful thinking. For a vision to be valid, it must be tangible and therefore measurable. It must also be in the future tense because it is really an aspiration. It is where we want the organisation to be at some determined time in the future.

Creating a good statement of vision is not that difficult, but it does require a lot of thought. All too many vision statements we see today are either copies of ones seen somewhere else just to look 'with it'. Or they are simply badly thought out because the authors did not know how to do it and either had no advice or were badly advised. This is unfortunate because the vision is probably the most important statement the organisation can make for its people, its customers and suppliers to know where the organisation is headed and its degree of determination to get there. Clearly, this needs careful thought.

In this chapter we only cover the rudiments of *Hoshin Kanri* and do not go into detail on how the Vision or any of the other attributes of the concept are created. Briefly, The vision should be created by the chief executive assisted by *all* his direct reports. For it to work they will all have to feel passionate about achieving its goals.

As mentioned earlier. This statement must be aspirational. It must give a clear indication of where the organisation intends to be headed and ideally, when it hopes to get there. It can have dates on it, but this is unusual.

We shy away from giving too much in the way of examples, mainly because we do not want to put words in people's mouths. Typically, top teams use what is often referred to as 'imagineering' to create them. A popular approach is for the team to use Post-it notes. Each member of the team is given, say, ten Post-it notes. Then they are asked to close their eyes and imagine that it is now five or maybe ten years into the future. What they must ask themselves is:

Imagine a journalist came to us from, say, *The Economist*, or the *Financial Times* or some other prestigious journal and we allowed them free access to all personnel, all departments, etc. including the supply chain and customers. A month or so later a report on our organisation was published across the whole centrespread. What would we like it to say? You must be specific. Imagine the banner headline. What would you like that to say? Now, using the Post-it notes, write down just one idea on each sticker that comes to mind.

When the members of the team have done this, then each of them posts their stickers on the wall but they cluster them according to type. For example, all those related to customers are in one group, suppliers in another, employees in another and so forth. Typically, there will be up to eight specific categories as shown in Figure 5.5.

When the members of the team have done this, then each of them posts their stickers on the wall but they cluster them according to type. For example, all those related to customers are in one group, suppliers in another, employees in another and so forth. Typically, there will be up to eight specific categories as shown on the next slide. The typical categories, which can vary, are:

- Customer
- Employee
- Suppliers
- Processes
- Organisation
- Technology
- Finance
- Design/development.

From the content of the relevant Post-it stickers, a relevant stamen of intent can be created for each of these.

Then from the complete group collectively, a statement can be created that summarises the overall direction this is intended to take the company, and this becomes the 'vision'. It is the policies that relate to each of the eight on the previous slide that collectively indicate how this vision is to be achieved. Just one final point. We said earlier that these aspirations had to be both tangible and measurable. Typically, we would study the Post-its and underline each of the key words on all the stickers that fit this requirement as it is these that lead to the creation of the key performance indicators.

The statements for each of the categories, which we will refer to as the 'drivers', collectively become known as the 'mission statement'.

In your work or home life

Away from work, why not get your family or some friends together, give them all some Post-it stickers and create a vision for next year's holiday!

Summary of learning

What you learned in this section:

- Who creates the vision and mission statements and how they are created
- How vision statements are created
- The identification of the drivers that will lead to the creation of key performance indicators.

SECTION 4: KEY PERFORMANCE INDICATORS (REFER ALSO TO SECTION 10, CHAPTER 3)

In the previous section we saw how key tangible success criteria could be identified from the process of creating the vision. In Figure 5.5 we will see how one of these mission statements, in this case the one created for the employees, was broken down into specific performance indicators. Note how in the statement the key tangible words are underlined, then these words are then copied into the second column of the tree diagram to the right. These are then broken down into the specific measures that relate to each.

The same process is carried out for the other seven criteria, customer, supplier, etc. So, we can see that just for this one element, it has been broken down into 22 specific measures, each of which is measurable.

The next step is to attempt to identify the current performance for each of the 22 items. What we will probably find is that we have some data on some of them but probably none on the others. Figure 5.5 shows a typical example of the measures for the customer driver. What is required now is to determine the current performance for each of these but, before we do there are 20 items on this list, there were 22 on the employee list and if it is similar for the other 6 then we are looking at some 160 items that we need data on. We have seen this list at over 400 in one case. Do we need all of that data? So, to determine whether or not we believe we are customer focused, the following are the issues we believe we need data on.

1. Number of customer suggestions implemented
2. Customer needs analysis
3. Customer exposure throughout the organization

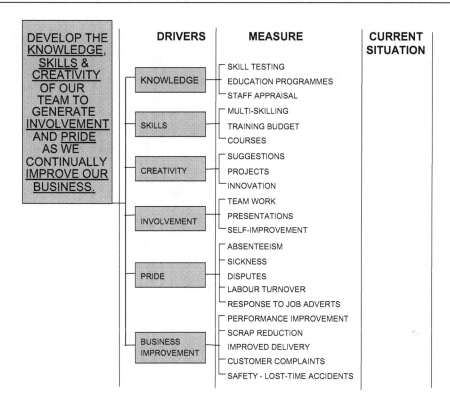

Figure 5.5   Tree diagram

4.   Build-to-order time and delivery time
5.   Number of customer complaints
6.   Customer praise
7.   Customer expectations vs. perceptions
8.   Percentage of calls transferred or passed up the organisation
9.   Average weekly calls per service area
10. Rate of parts outage
11. Number of on-time deliveries
12. Number of returns
13. Price relative to competitive market price
14. Customer willingness to pay for products
15. Extent products are more user friendly compared to competition
16. Number of repeat complaint calls
17. Positive press index
18. Negative press index
19. Dealer satisfaction index
20. Quality ratings
21. Warranty costs.

We can cut down the work considerably if we are careful. The reason being that not all the items will be equally important. There are some that we believe to be of great importance and others that would be nice to achieve but we think will only have a more modest impact on the achievement of the vision aspirations. So, the next step will be to rank them in perceived order of importance. In Chapter 4 we saw a ranking technique called paired rankings. However, we can get away with a simple vote if we ask our team to list them in order of importance. Not everyone will necessarily agree the precise order but usually it works. What we are hoping for is a list of some four or five items that if we do not achieve them, we have no hope in achieving our aspirations in the vision. The following might have been the opinion of the team after they had ranked the top ten of the list of 21 in their perceived order of importance.

The team might have used a simple vote or if they had wanted to be more precise and gain a greater consensus then they could have used the paired ranking technique, but it would have taken some time using this method to rank 21 items.

It is important to note that the rankings the team has agreed might represent market needs at this time but this can and will change over time. For example, suppose the team finds that perhaps a new competitor has entered the market and is clearly impacting on our market share. We conduct a study and find out that, contrary to our earlier opinion, price competition is far more important than we had previously thought, and the new competitor is exploiting our weakness. Then we re-order the rankings and this fluidity continues indefinitely. This way we can quickly react to any negative threats to our business.

1. Number of on-time deliveries.
2. Customer praise.
3. Build-to-order time and delivery time.
4. Number of customer suggestions implemented.
5. Customer needs analysis.
6. Number of customer complaints.
7. Quality ratings.
8. Number of returns.
9. Rate of parts outage.
10. Average weekly calls per service area

Customer exposure throughout the organisation.
Customer expectations vs perceptions.
% calls transferred or passed up the organization.
Price relative to competitive market price.
Customer willingness to pay for products.
Extent products are more user friendly compared to competition.
Number of repeat complaint calls.
Positive press index.
Negative press index.
Dealer satisfaction index.
Warranty costs.

Now that we have a shorter list, we then need to set targets for each. When we have done this, we then have our key performance indicators. What we have agreed is that these are critical to our success. In fact, some organisations refer to them not as key performance indicators but critical success criteria. There are also other terms in popular use, but they all have the same meaning.

Here is an example of the use of KPIs from a Japanese refrigerator factory. The number in the top left corner **520** is a KPI and represents the number of refrigerators targeted for that day's production.

In Figure 5.6, to the right the number **273** is the number of refrigerators that should have been produced at that moment in the day if the target is to be reached. The number below it is the number that have actually been produced. This number will fluctuate. Finally bottom left, the **13** represents the number of minutes lost since the start of the shift due to interruptions. In Figure 5.7 we see what can be done about that.

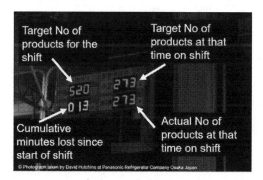

Figure 5.6   **ANDON explanation – 1**

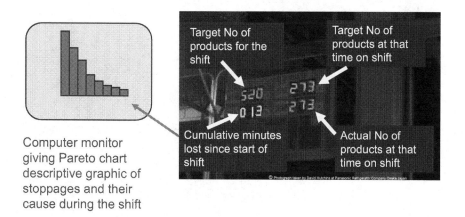

Computer monitor giving Pareto chart descriptive graphic of stoppages and their cause during the shift

Figure 5.7   **ANDON explanation – 2**

In your workplace

Think about how the use of key performance indicators would help in your work or possibly also your homelife.

Summary of learning

What you have learned in this section:

- What are tree diagrams and how they are used to derive drivers and measures from the mission statements
- You learned how performance indicators are derived from the measures and how to create key performance indicators from the performance indicators.

SECTION 5: DEPLOYING THE VISION TO ALL LEVELS

What you will learn in this section:

- The criteria required to successfully deploy the vision and mission elements down through the organisation
- How to use the deployed key performance indicators for improvement plans

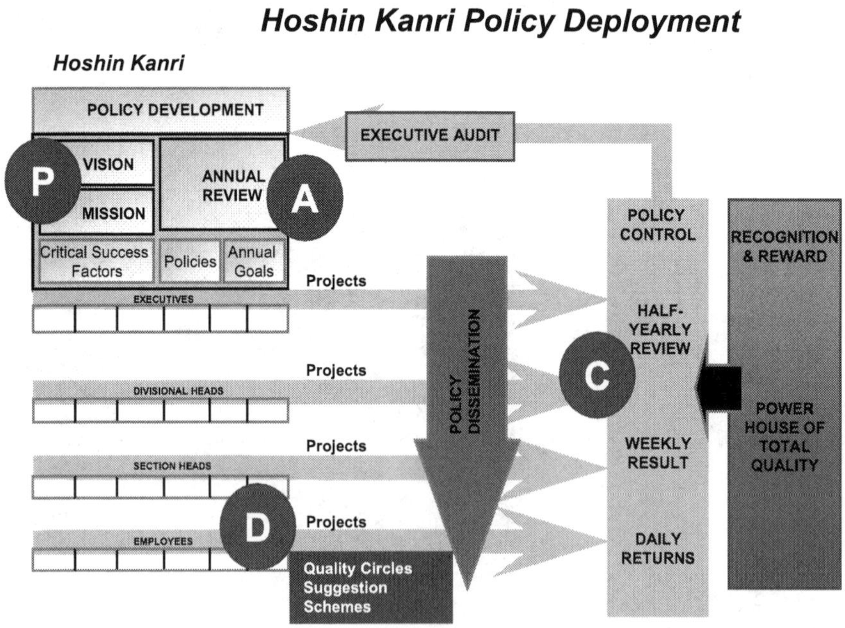

Figure 5.8   *Hoshin* deployment model

For *Hoshin Kanri* to work effectively, it **must** be deployed down through the organisation. This is from the top, layer by layer and no layer should be overlooked. At each layer, the vision through to KPIs is shared but then each layer will be encouraged to create a vision of its own, which, as we have said, must be compatible with the one above. This continues right down to the level of the individual employees. This is not dissimilar to the process adopted by some staff appraisal approaches. This is often referred to as a catch ball process.

This diagram, which for clarity is repeated but enlarged in Figure 5.8, shows the complete *Hoshin* model. In this version we have used the term 'critical success factors' instead of KPIs, but the terms are interchangeable. Note that the PDCA cycle that we discussed earlier is clearly present. In this instance:

- The P represents the plan.
- The D represents the deployment to all levels (Do)
- The C is the checking or auditing process, and
- A is the annual analysis of the data from the audits to set new KPIs for the following year.

Figure 5.9 shows how the cascading KPIs can then be converted into targets and displayed graphically to chart progress from current performance towards the

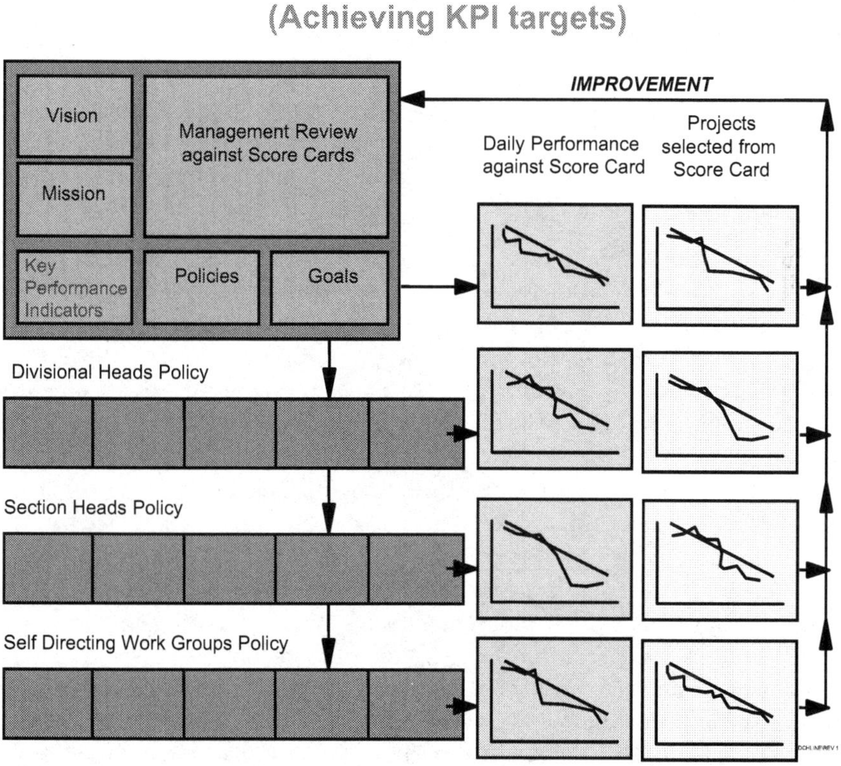

Figure 5.9   Flagship approach

achievement of the KPIs. Komatsu in Japan refer to this as their 'flagship' approach – an analogy using the fishing boats' use of flags to show the size of their daily catch.

Pause for thought

Think about how you might use the flagship approach in your work.

Summary of learning

What you learned in this section:

- How to construct a model that is based on the PDCA cycle to successfully deploy the vision and mission elements down through the organisation
- How to use the deployed key performance indicators for improvement plans.

SECTION 6: THE HOUSE OF QUALITY

What you will learn in this section:

- What is the 'house of quality' and how it is used in *Hoshin Kanri*
- How the KPIs derived from the vision and mission statements can be used to identify who is involved in their achievement
- How, using a central matrix all goals, current performance and organisational responsibilities can be displayed on a single graphic.

The house of quality is a brilliant concept, which was originated in 1972 at Mitsubishi's Kobe shipyard site. Initially, it was part of a concept known as quality function deployment but quickly also became an integral part of *Hoshin Kanri* management systems. In this section we will show how the KPIs created from the vision and mission statements can by using the house of quality concept can be organised in such a way to become part of the quality improvement process we looked at in the previous chapter.

Recall that we went from each of the possibly eight mission statements, through highlighting the tangible words (in our example, 'first choice', 'stated needs', etc.) as 'drivers' to the identification of the measures for each, to a determination of the current situation as shown here. From this we identify for each from benchmarking what is the desired performance, which we called the KPI for that measure. What we need to do now is to determine who should be responsible for the achievement of each specific KPI and we can do it in the following manner. For each specific KPI we can first identify all the functions in the organisation are directly involved in its achievement. One way of doing this is to use the process mapping tool that we looked at in the last section. This will enable us to determine each of the functions or activities which are directly involved.

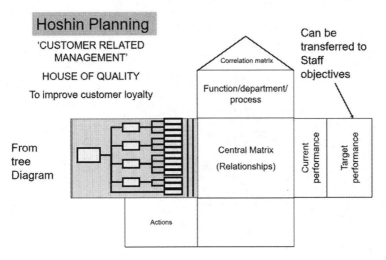

Figure 5.10   House of quality

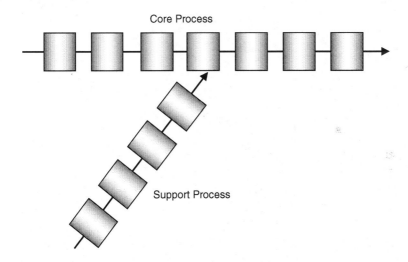

Figure 5.11   Support processes

We could create process flow diagrams for all KPIs, but the result would be very messy. The house of quality concept allows us to do this for all KPIs on one single diagram where we can at a glance see everything that we might need to know. We can do this by the following method. In Step 1, first of all, we take the tree diagram that we created for each mission element and place it on the left-hand side of the central matrix as shown in Figure 5.12.

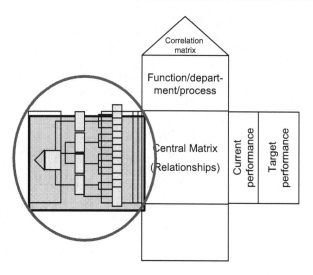

Figure 5.12   House of quality assembly – 1

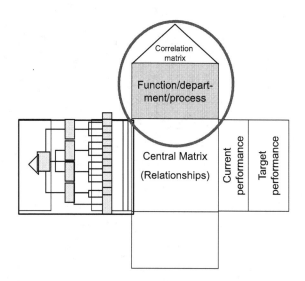

Figure 5.13   House of quality assembly – 2

In Step 2, we Identify each of the functions/activities that exist within the organisation. Ideally, this should also include the supply chain but this would make the diagram too complex. In order to simplify the diagram we can include the supply chain manager and then for that function specifically create a sub-diagram or even multiple sub-diagrams for each class of supplier where the supply chain is complex as in the case of automobile or aircraft manufacture.

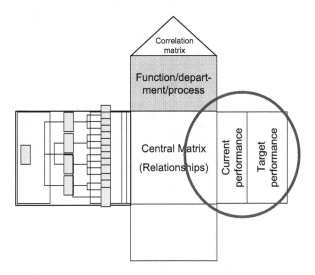

Figure 5.14   House of quality assembly – 3

In Step 3, we then add on the right-hand side the data we have computed for current and target performance but as we will see in the next section we also prioritise these in order of importance. There just remains the 'roof' of the house, which is the correlation matrix, but this is not so important when we are using the house of quality as part of *Hoshin Kanri*. It is more important to a concept known as QFD (quality function deployment), which we deal with later.

Summary of learning

What you learned in this section:

- How the concept of the 'house of quality' is constructed and how it is used in *Hoshin Kanri*
- How the KPIs derived from the vision and mission statements can be used to identify who is involved in their achievement
- How, using a central matrix all goals, current performance and organisational responsibilities can be displayed on a single graphic.

SECTION 7: MANAGING THE IMPROVEMENT PROCESS

What you will learn in this section:

- How the house of quality is used to organise the improvement process related to KPIs in need of attention.

- How responsibilities are identified
- How the use of symbols identifies who is responsible for what related to specific activities.
- How to put the entire organisation into a state of continuous improvement.

When complete this part of the diagram will look something Figure 5.17. Now that we have a matrix with all of the departments or activities across the top and all of the specific KPIs vertically to the left, we can now begin to identify the roles of each activity for the achievement of each KPI.

To illustrate this let us see how it works just for the first of our KPIs – 'on-time delivery' (See Figure 5.16). If then a problem arises in this process or it gets behind on its plan to achieve the KPI then a project team led by planning and production control and involving personnel from engineering services, production, transportation and packing would then use the Improvement process to solve the problem or to get back on track.

Now let us look at the right-hand side of the chart in Figure 5.17. We can see that for each KPI on the left-hand side, the specific 'measure' is stated on the right.

To the right of that follows a column stating the importance of the parameter, then level of customer expectations, the current level achieved and

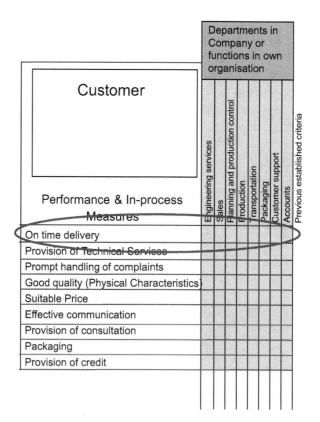

Figure 5.15   KPI analysis

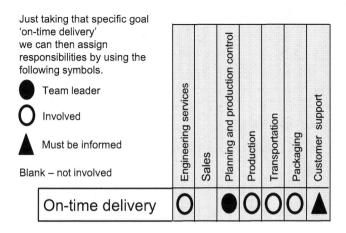

Figure 5.16   Responsibilities

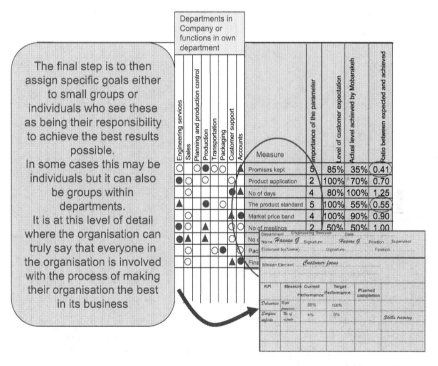

Figure 5.17   Assigning tasks

finally the ration between the expected or desired value and the achieved value. In this example, it was the policy of the company that every time the ration fell below 80% for a Level 5 priority item it automatically triggered an improvement project.

Pause for thought

Think about some of the problems in your work and if a team is formed to solve some of them, who might be the best be?

Summary of learning

What you learned in this section:

- How the house of quality is used to organise the formal use of the improvement process related to KPIs in need of attention
- How individual responsibilities are identified, and personnel selected for project teams concerned with the key business performance requirements
- How the use of symbols conveniently identifies who is responsible for what related to specific activities
- Finally, how well organised *Hoshin Kanri* planning puts the entire organisation into a state of continuous improvement.

SECTION 8: MANAGING RISK

What you will learn in this section:

- Where the work we have done in this book fits in to the comprehensive *Hoshin* model.
- How risk is identified and managed.

In chapter to this point, we have covered all the *Hoshin* model boxes on the top part of the diagram. We will finish the chapter with a brief look at the boxes in the lower part. Those marked in grey are designed for specific groups of people but the method of working is essentially the same.

In our chapter this far, apart from looking at product recall in the first chapter (because it can impact on everybody at random) we have focussed mainly on the most important concepts in quality that will enable us to become better at what we do. However, as we all know, life does not always run so smoothly. Not only do things go wrong unexpectedly but sometimes there are forces at work in the market place that are deliberately trying to undermine our efforts. Mostly, this particular threat comes from competitors but not always. Here we will take a brief look at other forms of risk.

Risks come in all shapes and sizes. Some are predictable, and others are not. We can deal with risk in four possible ways.

1. Ignore it and hope that it does not happen.
2. Insure against it 'IF' we can find someone to cover it and if we can afford the premiums.

3.   Self-insure and keep a fund available to cover it when things go wrong.

4.   Apply rigorous quality strategies to reduce the risks to a minimum.

Let us look at each of these. They are not all mutually exclusive as we will see.

1. Ignore it and hope that it does not happen

Believe it or not, some organisations make a conscious decision to do this. They may have investigated the risks, the costs of insurance and decided that if the risk is realised and that they cannot afford the consequences, they will simply go out of business. However, they can still make this less likely to happen by implementing a programme such as we are discussing here.

2. Insure against it 'IF' we can find someone to cover it and if we can afford the premiums

This is often referred to as 'passing the risk to others'. It is OK if you can find someone to underwrite it but as litigation becomes more common and as penalties for quality and safety-related failures increase. Increasingly common is the approach where insurance is possible up to a fixed sum. Once that amount has been paid out then the organisation is no longer insured. To repeat the insurance, *if* the company can find an underwriter, will usually incur a significantly greater fee but more likely, the risk will not be accepted.

3. Self-Insure and keep a fund available to cover the risk when or if things go wrong

This is little different from no insurance except that funds are accumulated to offset the costs should the worst happen. This is a sensible strategy to run alongside insurance anyway to provide some cover should the insurance company refuse further cover or if the resultant litigation costs more than the insurance limit, which is often a fixed amount with this type of insurance.

4. Apply rigorous quality strategies to reduce the risks to a minimum

This makes a lot of sense whichever alternative of 1, 2, or 3 is adopted. Today, many insurance companies are beginning to link premiums to the existence or not of well-structured quality management systems. A good quality track record can have many hidden compensations such as this.

Pause for thought

Think about some of the issues in your work that could put the organisation at risk.

Summary of learning

What you learned in this section:

* Where the work we have done in the first five chapters fits in to the comprehensive *Hoshin* model and you learned how *Hoshin* planning is effective in enabling the strategic issues to be organised.

> • How risk, which is also featured in the current version of ISO 9000, is identified and managed.

SECTION 9: COMPETITIVE ADVANTAGE

What you will learn in this section:

- • How serious competitors can look for and exploits our weaknesses.
- • How we can exploit the weaknesses of others through the use of the 'loose brick' concept.

The loose brick!

We have included this short section close to the section on risk because the two are directly related. Those who are concerned with business strategy might view the David Hutchins article on the DHI website (www.dhiqc.com) entitled 'The loose brick'. This is a term used in Japan where a sharp-eyed competitor will search for a weakness in their competitors organisation, which it can exploit in the market place.

The loose brick concept is well known in Japan. The nearest term we have in English is 'The Achilles heel'. Back in 1974, in the UK there were over ten UK and European TV producers. These included: McMichael, Bush, Decca, Ferguson, Grundig, HMV, Philips, Rediffusion, Sobell, Ultra, etc. Just one year later almost all of them had disappeared as TV producers. Why?

In the high street store windows, these had all been replaced with Sony, Toshiba, JVC, Hitachi, Sharp and Panasonic, etc. Names we had never heard of before. Philips carried out a defensive study to find out what was so special about the Japanese TVs that was having such a catastrophic impact on everyone's market share. They found the reason.

Figure 5.18   Competitive advantage

In those days it was an industry benchmark that a new model of TV had to be reliable to the extent of approximately 6% of them requiring a maintenance call (or worse) during the first 12 months.

This was a well-known statistic in the trade and the dealerships were sensitive about it not wanting to have their repair shops overloaded with defective TVs. Philips found that the Japanese sets only failed at the rate of 0.5% (later it fell to 0.1%). This was an unheard-of statistic in the West. This was the loose brick!

The Japanese manufacturers had studied the market, found the sensitivity towards the 6% figure, then stayed out until they were so far ahead that the Western competitors could not catch up in time to survive. In most cases that was correct, but they had not taken Philips' smart observation into account. The CEO of Philips sent an edict round to all his suppliers. To get down to the 0.5% meant shipping parts at the defect rate of parts per million (ppm) instead of current practice of parts per 100. The suppliers had no way of knowing how to achieve this, but Philips said that unless we achieve it within a year, we are dead. If you want to survive alongside us, then you will have to allow our engineers to come to your plants and work alongside you to achieve that level of performance. That is what happened and that is why Philips survives to this day. They were the only competitor who recognised the threat. Ferguson also survived but that was for a different reason. They had a niche market in small B&W TVs useful in Caravans. At that time nobody else was in that market.

Today, the Japanese do not say so much about the loose brick but instead have another concept they call *Dantotsu*. The word means 'number one'. It does not just mean being regarded as the 'number one' company but it means 'number one' for every single product that it makes. Not only when the product is launched but for an estimated five years after. Komatsu claims to have achieved this for over 80 of its 150 product types.

There are three aspects to the loose brick idea.

1. We need to be sure that our competitor has not already discovered a loose brick of ours and is working towards demolishing us by removing it from his own performance record and then demolishing us.
2. Are we looking for a loose brick in our competitor that we think we can usefully exploit?
3. Is there a loose brick situation where both we and our competitor are vulnerable to a currently non-competitor?

The non-competitor (so far!)

At one time, lighting was by either candle or oil lamp. Today, it is predominantly from the electric light. At one time students used mathematical tables and a device called a 'slide rule' to do calculations. In both above now almost obsolescent industries, none of the companies involved could have imagined that they were about to be crashed out of business. The competition came not from one of their traditional competitors but from a totally different industry that destroyed the need for their products at a stroke. Are you so vulnerable?

Pause for thought

Can you see any possible 'loose bricks' in your organisation' now? Also, can you see any exploitable 'loose bricks' in your competitor's strategy?

Summary of learning

What you have learned in this section:

- How the 'risk' concept can be used not only to reduce the risk of negative impacts on our organisation but also how we can use it to exploit the weaknesses of our competitors.
- How your organisation can make this exploitation of the weaknesses of competitors using the 'loose brick' concept.
- Also, how some of the biggest threats can often come from organisations that are currently in completely different fields. The graphic example used was the elimination of oil lamp manufacturers by manufacturers of electrical product with the advent of the light bulb.

SECTION 10: LINKING TO STANDARDS AND AWARDS

What you will learn in this section:

- The effect of standards on variety reduction
- The role of standards in the improvement of safety.

In favour of standards

One of the most compelling arguments in favour of product standards is the reduction in unnecessary variety. We will not go into the history of standards here, but briefly, the standardisation industry has its roots in UK history when at the turn of the twentieth century, the engineer Sir John Wolfe Barry made note of the fact that at that time there were over 70 different designs of railway track in use. This led to enormous unnecessary costs. By a process of simplification, his work resulted in a reduction to just five. This gave rise to huge economies of scale and the standards industry was born. Therefore, there are today literally thousands of standards for almost everything.

Whether we generally like the idea of standards, regulations, specifications or not, the fact remains that we could not do without them. If you live close to a nuclear power station, you would not feel very comfortable if you suspected that no well thought out, and rigorously tested, procedures were standardised and religiously carried out. Also, you would not feel very secure if you thought that the

Figure 5.19   Awards encourage pride

application of these tested methods was left to chance and that there were not frequent inspections and audits to ensure that all is well. However, you know that the application of such rigid systems is expensive, and you would not expect the same level of rigour in less life-threatening situations. The expression is 'horses for courses'.

Fortunately, the standard ISO 9000 has been designed to allow for this flexibility. It is generic and therefore applicable across industry but it by no means follows that all you must do is to apply ISO 9000 and hey presto all your quality problems magically disappear. Sadly, this is how it is being sold by some and the standard lacks credibility in many people's eyes for that reason. What ISO 9000 does is to provide a base to work from. It gives those who do not have a strong foundation in the quality-related arts and sciences an opportunity to create some basics on which to build. If you think about it this should be obvious. You cannot standardise on being the best. Even in a situation where several competitors all had ISO 9000 certification, they could still be ranked in terms of low cost to produce, customer satisfaction, etc.

If we study the reasons why, even when working from a common base, still some organisations are significantly better than others, we find that these organisations go far beyond the parameters of any standard. This is the beauty of *Hoshin Kanri*. *Hoshin* is developed from the needs of the organisation and includes everything that will make it successful if carefully designed and implemented. If a *Hoshin* program is well designed then not only will it automatically contain all the requirements of ISO 9000, but it will embrace all the requirements of alternative standards, product standards, test standards and the various quality award programmes. Of course, this needs testing, so we mention the main ones here.

Before we do, we need to separate standards into their various types. So far, we have mentioned systems standards (ISO 9000) but there are industry specific versions of these especially for the aircraft and automotive industries. PS 9000 is the pharmaceutical equivalent of the automotive QS 9000 and like the automotive standard there have been updates. There is a new ISO standard ISO 15378, which is a risk-based GMP (good manufacturing practice) standard that incorporates several requirements, in particular additional documented procedures, additional to those in PS 9000. The GMP principles in the two standards are similar; it is primarily the detail that varies.

In the food industry there is the ISO 22000 series. This is an international standard, however, there are in the region of 20 food safety standards throughout the world. In the United Kingdom there is the British Retail Consortium (BRC) Global Standard for Food Safety. This standard has been produced in reply to several large supermarkets requesting a common standard. In Germany and France there is the IFS (International Food Standard), again produced in conjunction with their food trade associations.

To ensure compliance with these standards, there is a global third-party audit system. Administration of this system varies from region to region and industry to industry but the principles of all of them are similar.

Of relevance, there are also product test standards such as those related to the BSI Kitemark and CE marking. There are standards for auditing procedure, sampling methods (BS 6001 Series) and even for training programmes.

The practice of auditing is not covered in this book. The reason being that it is the most widely used and known aspect of quality worldwide. Both the internet and textbooks are saturated with explanations of auditing technique and this book is more concerned with philosophy.

Having standards for all of these concepts has its advantages. The main one being, as previously stated, it provides consistency based on what is perceived to be current best practice (at least in the eyes of those who set the standard). But it is by no means a perfect solution. Flaws include:

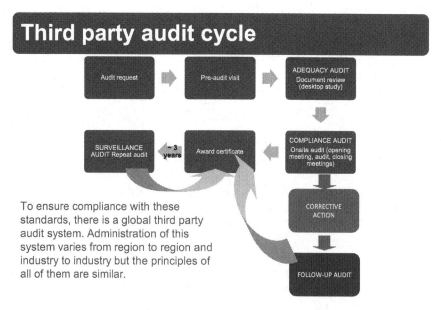

Figure 5.20   The audit cycle

- The risk of standardising on something in such a way as to give some an unfair advantage.
- It inhibits innovation. Once standardised, a method is difficult to change both due to cost and to political expediency.

A good example is the wall socket. There are three principal designs in use in the world today. The one used in the UK and the old Commonwealth, that used throughout most if not all of Europe and the one used in the USA and countries heavily under its influence. Now it is likely that one of these must be better than the other two, but the world will never change. Why? Because none will want to give an economic advantage to the other during the change process. There has been an ISO committee meeting regularly on this topic now for several decades!

SECTION 10: LINKING TO STANDARDS FOR MANAGEMENT SYSTEMS AND NATIONAL AND INTERNATIONAL QUALITY AWARD PROGRAMMES

Probably one of the best known of these is the Investors in People scheme. It is similar to a standard in as much that certain criteria are examined or audited It is a standard of good practice in developing people and is intended to be achievable by all organisations. It enables organisations to benchmark their own practices against a nationally recognised standard.

The EFQM Excellence Model (European Organisation for Quality Management Excellence Model)

This was launched in 1992 and thought to be a response to the earlier success of the American National Quality Award known as the Baldridge Award.

The theory behind the award is based on the premise that organisations, regardless of sector, size, structure or maturity, need to establish an appropriate management framework to be successful. is a practical, non-prescriptive framework that enables organisations to:

- Understand their key strengths and potential gaps in relation to their stated vision and mission
- Effectively communicate ideas within and outside the organisation
- Integrate existing and planned initiatives, removing duplication and identifying gaps
- Provide a basic structure for the organisation's management system.

The underlying principles of the award, which it claims are the essential foundation of achieving sustainable excellence for any organisation, are:

- Leadership
- Policy and strategy
- People
- Partnerships and resources

- Processes
- Customer results
- People results
- Society results
- Key performance results.

Limitations

Although there is an option for 'scoring' and an awards recognition scheme, they may be expensive for smaller social enterprises to enter. There is no formal mark or accreditation for the model and it will not be overtly visible or recognisable to customers, service users, funders and other stakeholders. However, EFQM does offer awards to organisations, showcasing winners through its website and networks. Though it has been used successfully by several medium and large voluntary organisations, it was initially developed for the commercial sector and some of the language of the *excellence model* may not translate easily to social enterprises or voluntary organisations. There are limited examples of use by not-for-profit and non-governmental organisations.

Other international awards

In the USA the National Quality Award, is referred to as 'The Baldrige Award', so named after one of the members of the organising committee was killed in a rodeo accident. This was launched in the late 1980s, largely as a response to the success of the Japanese Deming Prize but unlike the Deming Prize, it has a specific model, which does not look like the EFQM model which works in much the same way. In Japan they have the Deming Prize, which dates to the late 1940s. Whilst it carried Dr Deming's name, the award preceded his involvement and was originally based on auditing practice. His name was added as a gesture of appreciation for a series of lectures that he gave in 1950. There is no model for the Deming Prize as it is given for the application of *Hoshin Kanri* and in this case every organisation would have a different-looking model that is self-created.

Pause for thought

Think about some of the standards that might be used in your work.

Summary of learning

What you have learned in this section:

- How standards can be used to achieve variety reduction, the adoption of what are currently best practices but that they could stifle innovation if not properly managed
- Where standards and standardised management systems and test methods can instil confidence
- The role of standards in the improvement of safety.

---

6 Quality-related legislation and regulation

WHAT YOU WILL LEARN

- Legal definition of a defect in statute law and its relevance to the design, manufacture and distribution of goods and services
- Legal framework – UK, EU
- The principles of common and statute law
- The role of precedent and case law in the law of tort
- Common law applied to quality. The common law relationship between buyer, seller and third parties
- Statutory law and its impact on the responsibilities of an organisation and quality professionals
- Common statutory requirements
- Sale of Goods legislation
- Consumer protection legislation
- EU directive on product liability
- Occupational safety and health.

CAVEAT

This chapter includes details of some relevant laws for descriptive purposes only and it must be pointed out that this is their sole purpose because law is a living organism. Not only can the clauses in these laws be changed even whilst this book is being written, but the interpretation of those laws can be changed almost at any time because of case law as well. Our advice to all readers is that if you are in anyway concerned that you are at risk or might be a potentially injured party with regards anything included in this Chapter to both seek professional legal advice before doing anything and to check the current version of the laws in question as these are readily accessible mostly on line. Remember! Ignorance of the law is not a valid defence.

- Section 1: introduction and basic legal principles of civil and industrial law
- Section 2: development of case law related to quality
- Section 3: some important quality-related cases
- Section 4: products liability – the EU directive
- Section 5: products liability – the UK Consumer Protection Act
- Section 6: statute and common law related to weights and measures
- Section 7: Occupational Safety and Health (OSH)
- Section 8: product certification schemes
- Section 9: risk assessment.

SECTION 1: INTRODUCTION AND BASIC LEGAL PRINCIPLES OF CIVIL AND INDUSTRIAL LAW

The principles and content of both civil and industrial law vary across the world from country to country and in many cases from state to state as is the case in the USA. However, there are some commonalities. The main one is the principle of judicial precedent, which we will come to later.

The two largest groups are the laws that developed in Western democracies and Muslim law, known as Sharia law. Students from outside the UK would do well to study the laws as they apply in their own countries and if concerned with exporting or if working for a multinational organisation, the laws in the specific countries that might be relevant.

Statute and common law related quality

During the early centuries of village life, consumer protection against poor quality was largely based on the principle of *caveat emptor* – let the buyer beware! In those days producer and consumer met face-to-face in the village marketplace. The goods on display consisted mostly of natural products, or products that had been produced from natural materials. Consumers had long familiarity with such goods.

Risks related to the law

As previously stated, ignorance is no excuse in law, so to reduce risk, be better informed. Having determined the risk, there are three main options open to an organisation in order to respond to possible litigation. There are others that will be mentioned in the section on 'risk assessment'.

1. Ignore it and hope that it does not happen and be prepared to be forced out of business if the risk is realised and you or your organisation is successfully prosecuted.
2. Transfer the risk to others if they are willing to take it through insurance.
3. Reduce the risk by having good quality-based management organisation in place.

Items 2 and 3 above are not mutually exclusive. In some cases, especially with regards product liability insurance, premiums can be related to the quality of organisation.

Origins of UK particularly England and Wales (as Scotland is a slightly different) civil and industrial law

A brief history is given below of some of the key points of UK law, which is also the root of much of the law around the world. Scottish law differs slightly from that of England and Wales.

The Magna Carta

The Magna Carta, which dates its origin to 20 June 1214, is the basis of much of the law around the globe. It is commemorated in a memorial site near to where it was signed by King John at Runnymede by the River Thames, UK, between Windsor and Staines. There are two main branches of Law: statute law and common law.

Statutory laws are those that result from Acts of Parliament such as the Consumer Protection Act, the Road Traffic Act, etc. Breaking these is a 'criminal offence'. Common law is also known as case law or the law of torts as they are based on court judgments resulting from disputes between individuals.

Criminal law

A proven breach of statute law is regarded as a criminal offence and is therefore also known as *criminal law. Depending upon the offence and its severity, this can result in jail, fines, suspended sentences or cautions.*

Common Law

Common law is different from statute law as it results from disputes between individuals and not between individuals and the state. The two main aspects to this type of Law are:

- Trespass
- Negligence

Trespass

This is the more difficult because **intention** has to be proven. Even where there are warning signs, the defendant can claim that they did not see the signs. A host of reasons can be given. However, if it can be proven that the defender was aware, then they will probably lose but if they leave the area in question immediately following the warning then they should be safe.

Negligence

This is the most important area of common law from a quality point of view. In this case, by definition, intention does not have to be proven. It comes up frequently in cases where there is a claimed violation of 'the duty of care' we owe to others. It is also included in some aspects of statute law for example, the Health and Safety at Work Act.

'All reasonably practical steps' should be taken to ensure that ...' etc. The Consumer Protection Legislation dates back to 1896. The original Sale of Goods Act required 'goods to be of merchantable quality and fit for the purpose intended'.

Judicial precedent

This originated with case law or common law in the centuries following the Magna Carta. In those early days there were virtually no statute laws as we know them today, and the laws, such as they were, resulted from decisions by the king's travelling judges being used to judge later cases that were similar in type.

This was known as judge-made law and the process of using previous decisions to decide current cases was called judicial precedent and applies in both statute and common law to this day in many countries.

Judicial precedent – testing through the courts

For example, suppose that Parliament introduced a new law related to, for example, alcoholism, where the content of the law was radically different from the previous law. The wording of the new law might in some cases be open to interpretation. It would be argued that any of the possible interpretations could be possible 'until that aspect of the law had been tested through the Courts'.

Judicial precedent – current situation

Once the courts have interpreted the meaning then that judgment will be used in subsequent cases subject to appeal when it can be reinterpreted for the current case and for the future. This applies to both common law and statute law.

Another early example from medieval times was a technique used in the building of timber framed houses. If one looks carefully at the ends of each piece of wood, in the authentic buildings there will usually be a craftsman's mark, which will be identical to a further mark on the adjoining piece of wood (Hutchins 1995, Dr Juran's *A History of Managing for Quality*, p. 434). The marking of craftmade products is still used to this day in the pottery industry in the case of high-quality china products where they are hand painted. In many cases, the craftsman who did the painting will put their own mark on the underside, which identifies them with the product.

SECTION 2: DEVELOPMENT OF LAW RELATED TO QUALITY

Before the Sale of Goods Act 1893

Before the Sale of Goods Act 1893, there was no specific legislation related to the quality of goods as such. One of the early examples of 'quality', which still impact our lives, are the routes of the old Roman roads. It is still not known for sure how they made them so straight. The most popular theory is, that they lit three fires in a straight line on the high ground, that were visible from the point the road had reached. As they progressed, they lit further fires ahead and so on.

What there was, was a succession of laws and regulations intended to protect the consumer against some of the activities of rogue traders, for example, the hallmarking of silver, the establishment of nationally approved measures known as the avoirdupois system or the Imperial system for example, the ton, hundredweight, miles, yards, feet and inches, etc. These varied at one time across the country. In fact, just for the gallon there were some 26 different measures. Unfortunately, the UK adopted one of these and the USA another!

Statute and common law related to consumer protection and health and safety

With the growth of industrialisation and commerce, along with the rise of towns and cities, the producer and consumer no longer met face-to-face. The flow of goods now passed through intermediate processors and merchants. Many critical qualities could no longer be judged by the unaided human senses. In the towns and cities, the village marketplace became largely obsolete, and with it the doctrine of 'caveat emptor'. In its place, there emerged laws to help protect the consumer against poor quality, and to reduce the widespread practice of cheating.

To a considerable degree, quality could be judged by the unaided human senses. Under the circumstances, the principle of *caveat emptor* was a sensible element of consumer protection. It demanded that consumers learn to protect themselves through vigilance and self-reliance. Nevertheless, there were added forms of consumer protection available in some towns. The Craft Guilds tended to exercise control over their members, and this control was especially strict with respect to quality. In addition, the town authorities tended to deal severely with flagrant abuses by tradesmen.

For example, in 1365 John Russell, a poulterer, was charged with exposing pigeons for sale that were 'putrid, rotten, stinking and abominable to the human race; to the scandal, contempt, and disgrace of the city'. He was sentenced to the pillory and the said pigeons burned under him.

Sale of Goods Act (1893)

Prior to 1893, most statute law enacted for consumer protection related to weights and measures. Then, under the Sale of Goods Act (1893), a contract of sale in the book of business normally implies that 'goods are of merchantable quality and fit for the purpose intended', that is, that they are reasonably fit for the purpose for which goods of that kind are usually bought. The effect is to impose strict liability. However, there are two limitations imposed by the principle of privity of contract ...

Under this principle, the shortcomings are:

1.   The remedy is available only to the purchaser.
2.   The remedy is available only against the seller.

Under these limitations, if a third party suffers any loss in any way, he/she cannot use the Sale of Goods Act as the basis of his argument. His remedy would have to come through common law based on tort. However, under tort the plaintiff must **prove** that the defendant was negligent in some way and that this negligence led directly to the loss that the plaintiff argues that he has suffered. An example of this, 'The case of the snail in a bottle', is illustrated in the next section.

To do so the plaintiff must prove that:

1. The product was defective.
2. The product caused the injury or loss.
3. The defendant failed in his duty of care because the injury or loss was a foreseeable consequence of the defect.

Despite the seeming protection, this was not the case, and prior to 1936 there had been no successful actions based on negligence. The reason was the extreme difficulty of proving negligence due to a lack of expert knowledge of the processes involved in the manufacture of the product in dispute, and to lack of access to the production process or any records pertaining thereto.

SECTION 3: SOME IMPORTANT QUALITY-RELATED CASES

The case of the snail in a bottle

This difficulty was well dramatised by the first successful action, namely the case of **Donoghue v. Stephenson (1932) AC 562, 1932 SC (HL) 31.** (This was a defining case.) This became an important reference in subsequent developments in laws relating to consumer protection. In this case a woman was injured as a result of drinking a bottled beverage containing a decomposing snail! The injury was clearly traceable to the manufacturer – the bottler of the beverage.

The woman was unable to secure redress under the Sale of Goods Act due to lack of privity (a friend of hers had been the purchaser). Then, despite the lack of precedent, she sued the manufacturer. The judge ruled in her favour, and the case was appealed all the way to the House of Lords, which upheld the judge's ruling by a majority of three to two.

The English Law Commission recommended a change in the burden of proof; that is, that rather than the plaintiff having to prove negligence, it would be the responsibility of the defendant to prove that he was not. This would, of course, make it considerably easier to obtain a remedy in common law situations, but it violates the principle that someone is innocent until proven guilty. The law now states that the producer of an article shall be liable for damage caused by a defect in the article, whether he could have known of the defect. The producer shall be liable even if the article could not have been regarded as defective in the light of the scientific and technological development at the time when the article was put into circulation. **In other words, if a product hurts you, then it is defective.**

SECTION 4: PRODUCTS LIABILITY – THE EU DIRECTIVE

Competitive advantage

The main concern in Europe regarding differences in product-related legislation is the effect on competitive advantage. Clearly, a country with the laxest laws in this respect has lower manufacturing costs due to lower insurance premiums, less litigation and the ability to get away with fewer controls on the manufacturing process. Of course, those who understand total quality know that this does not have to be the case, but that was the prevailing opinion of the day. **In the end, the EU directive on this with some modifications was signed by the British government.**

The EU directive – caveat

There follows extracts from the UK consumer protection legislation. Please note, in both cases these are only extracts included to bring out some of the main points of the law aiming to provide a general understanding of the principal concepts involved.

In your work, we reiterate that should any of this be relevant then we strongly advise that professional legal opinion be sought. Whilst the principles as explained here are easy to follow, as in every case, the detail is critically important and must be understood. This goes beyond its relevance to this book.

Product liability – the EU directive

Article 1

The producer shall be liable for damage caused by a defect in his product.

Article 2

For this Directive 'product' means all movables, with the exception of primary agricultural products and game, even though incorporated into another movable or into an immovable. 'Primary agricultural products' means the products of the soil, of stock farming and of fisheries, excluding products that have undergone initial processing. 'Product' includes electricity.

Article 3

1. 'Producer' means the manufacturer of a finished product, the producer of any raw material or the manufacturer of a component part and any person who, by putting his name, trademark, or other distinguishing feature on the product presents himself as the producer.
2. Without prejudice to the liability of the producer, any person who imports into the European Community a product for sale, hire, leasing or any form

of distribution in the book of his business shall be deemed to be a producer within the meaning of this Directive and shall be responsible as a producer

3.   Where the producer of the product cannot be identified, each supplier of the product shall be treated as its producer unless he informs the injured person, within a reasonable time, or the identity of the producer or of the person who supplied him with the product. The same shall apply, in the case on an imported product, if this product does not indicate the identity of the importer referred to in paragraph 2, even if the name of the producer is indicated.

Article 4

The injured person shall be required to prove the damage, the defect and the casual relationship between defect and damage.

Article 5

Where, because of the provisions of this Directive, two or more persons are liable for the same damage, they shall be liable jointly and severally, without prejudice to the provisions of national law concerning the rights of contribution or recourse.

Article 6

1.   A product is defective when it does not provide the safety that a person is entitled to expect, taking all circumstances into account including:
     (a) the presentation of the product
     (b) the use to which it could reasonably be expected that the product would be put
     (c) the time when the product was put into circulation
2.   A product shall not be considered defective for the sole reason that a better product is subsequently put into circulation.

Article 7

The producer shall not be liable because of this directive if he proves:

(a) that he did not put the product into circulation; or

The producer shall not be liable as a result of this directive if he proves:

(b) that, having regard to the circumstances, it is probable that the defect that caused the damage did not exist at the time when the product was put into circulation by him or that this defect came into being afterwards; or
(c) that the product was neither manufactured by him for sale or any other form of distribution for economic purposes nor manufactured or distributed by him in the book of his business; or

(d) that the defect is due to compliance of the product with mandatory regulations issued by the public authorities; or

(e) that the state of scientific and technical knowledge at the time when he put the product into circulation was not such as to enable the existence of the defect to be discovered; or

(f) in the case of the manufacturer of a component, that the defect is attributable to the design of the product in which the component has been fitted or to the instructions given by the manufacturer of the product.

Article 8

1. Without prejudice to the provisions of national law concerning the right of contribution or rebook, the liability of the producer shall not be reduced when the damage is caused both by a defect in the product and by the act or omission of a third party.

2. The liability of the producer may be reduced or disallowed when, having regard to all the circumstances, the damage is caused both by a defect in the product and by the fault of the injured person or any person for whom the injured person is responsible.

Article 9

For the purpose of Article 1, 'damage' means:

(a) damage caused by death or personal injuries;

(b) damage to, destruction of, any item or property other than the defective products itself, with a lower threshold of 500 ECU, provided that the item of property
(i) is of a type ordinarily for private use or consumption and
(ii) was used by the injured person mainly for his own private use or consumption.

This article shall be without prejudice to national provisions relating to non-material damage.

Article 10

1. Member States shall provide in their legislation that a limitation period of three years shall apply to proceeding for the recovery of damages as provided for in this Directive. The limitation period shall begin to run from the day on which the plaintiff became aware, or should reasonably have become aware, of the damage, the defect and the identity of the producer.

2. The laws of Member States regulating suspension or interruption of the limitation period shall not be affected by this Directive.

Article 11

Member States shall provide in their legislation that the rights conferred upon the injured person pursuant to the Directive shall be extinguished upon the expiry of a period of ten years from the date on which the producer put into circulation the actual product which caused the damage, unless the injured person has in the meantime instituted proceedings against the producer.

Article 12

The liability of the producer arising from this Directive may not, in relation to the injured person, be limited or excluded by a provision limiting his liability or exempting him from liability.

Article 13

The Directive shall not affect any rights that an injured person may have according to the rules of the law of contractual of non-contractual liability or a special liability system existing now when this Directive is notified.

Article 14

This Directive shall not apply to injury or damage arising from nuclear accidents and covered by international conventions ratified by the Member States.

Article 15

1. Each Member State may
   (a) by way of derogation from Article 2, provide in its legislation that within the meaning of Article 1 of this Directive 'product' also means primary agricultural products and game;
   (b) by way of derogation from Article 7 (e), maintain or, subject to the procedure set out in paragraph 2 or this Article, provide in its legislation that the producer shall be liable even if he proves that the state of scientific and technical knowledge at the time when he put the product into circulation was not such as to enable the existence of a defect to be discovered.
   (c) A Member State wishing to introduce the measure specified in paragraph (1) shall communicate the text of the proposed measure to the Commission. The Commission shall inform the other Member States thereof.
   (d) The member state concerned shall hold the proposed measure in abeyance for nine months after the Commission is informed and provided that in the meantime the Commission has not submitted to the Council a proposal amending this Directive on the relevant matter. However, if within three months of receiving the said information, the Commission does not advise the Member State concerned that it intends submitting such a proposal to the Council, the Member State may take the proposed measure immediately.

2. If the Commission does submit to the Council such a proposal amending this Directive within the aforementioned nine months, the Member State concerned shall hold the proposed measure in abeyance for a further period of 18 months from the date on which the proposal is submitted.

3. Ten years after the date of notification of this Directive, the Commission shall submit to the Council a report on the effect that rulings by the courts as to the application of Article 7 (e) and of paragraph 1 (b) of this Article have on consumer protection and the functioning of the common market. In the light of this report, the Council, acting on a proposal from the Commission and pursuant to the terms of Article 100 of the Treaty, shall decide whether to repeal Article 7 (e).

Article 16

1. Any Member State may provide that a producer's total liability for damage resulting from a death or personal injury and caused by identical items with the same defect shall be limited to an amount that may not be less than 70 MECU.

2. Ten years after the notification of this Directive, the Commission shall submit to the Council a report on the effect on consumer protection and the functioning of the common market of the implementation of the financial limit on liability by those Member States that have used the option provided for in paragraph 1. In the light of this report, the Council, acting on a proposal from the Commission and pursuant to the terms of Article 100 of the Treaty, shall decide whether to repeal paragraph 1.

Article 17

This Directive shall not apply to products put into circulation before the date on which the provisions referred to in Article 19 enter into force.

Article 18

1. For the purposes of this Directive, the ECU shall be that defined by Regulation (EEC) No. 3180/78, as amended by Regulation (EEC) No. 2626/84. The equivalent in national currency shall initially be calculated at the rate obtaining on the date of adoption of this directive.

2. Every five years the Council, acting on a proposal from the Commission, shall examine and, if need be, revise the amounts in this Directive, in the light of economic and monetary trends in the Community.

Article 19

Member States shall bring into force, not later than three years from the date of notification of this Directive, the laws, regulations and administrative provisions necessary to comply with this Directive. They shall forthwith inform the Commission thereof (1).

Article 20

Member States shall communicate to the Commission the texts of the main provisions of national law that they subsequently adopt in the field governed by this Directive.

Article 21

Every five years the Commission shall present a report to the Council on the application of this Directive and, if necessary, shall submit appropriate proposals to it.

SECTION 5: PRODUCTS LIABILITY – THE UK CONSUMER PROTECTION ACT

Note! The following contains selected excerpts from the UK Consumer Protection Act. We recommend that should there be an issue related to this, that you first obtain a copy of the current version of the Act itself from HMSO and, if relevant, seek specialised legal advice. These notes are only intended to give an overall appreciation and the author is not a lawyer! By selecting relevant extracts from the Act itself, the numbering of paragraphs and sub paragraphs might not appear to be correct. As stated, please refer to the Act itself for detail.

Consumer Protection Act 1987: Part 1

An act to make provision with respect to the liability of persons for damage caused by domestic products; to consolidate with amendments with Consumer *Safety Act 1978* and the *Consumer Safety (Amendment) Act 1986*; to make provision with respect to the giving of price indications; to amend Part 1 of the *Health and Safety at Work etc. Act 1974* and sections 31 and 80 of the *Explosives Act 1875* and the *Fabrics (Misdescription) Act 1913*; and for connected purposes.

Part 1

Product liability

1. (1) This Part shall have effect for making such provision as is necessary to comply with the product liability Directive and shall be construed accordingly.
   (2) In this Part, except in so far as the context otherwise requires – 'agricultural produce' means any produce of the soil, of stock farming or of fisheries.

... 'producer', in relation to a product, means:

(a) the person who manufactured it;

(b) in the case of a substance which have not been manufactured but has been won or abstracted, the person who won or abstracted it:

in the case of a product which has not been manufactured, won or abstracted but essential characteristics of which are attributable to an industrial or other process having been carried out (for example, in relation to agricultural produce), the person who carried out that process.

Where two or more persons are liable by this Part for the same damage, their liability shall be joint and several.

2. (1) Subject to the following provisions of this Part, where any damage is caused *wholly or partly by defect in a product, every person to whom subsection (2) below applies shall be liable for the damage.*

(2) This subsection applies to:

(a) the producer of the product; any person who, by putting his name on the product or using a trademark or other distinguishing mark in relation to the product has held himself out to be the producer of the product; any person who has imported the product into Member States in order, in the book of any business of his, to supply it to another.

(3) Subject as aforesaid, where any damage is caused wholly by a defect in a product, any person who supplied the product (whether to the person who suffered the damage, to the producer of any product in which the product in question is comprised or to any other person) shall be liable for the damage

(4) Product means any goods or electricity and (subject to subsection (3) below) includes a product that is comprised in another product, whether by being a component part of raw material or otherwise; and

(5) The product liability Directive means the Directive of the Council of the European Communities dated 25th July 1985 (No. 85/374/EEC) on the approximation of the laws, regulations and administrative provisions of the Member States concerning liability for defective products.

(6) Subject to the following provisions of this section, there is a defect in a *product for the purposes of this Part if the safety of the product is not such as persons generally are entitled to expect; and for those purposes "safety" shall include safety in the context of risks of damage to property as well as in the context of risks of death or personal injury.*

This section shall be without prejudice to any liability arising apart from by this part.

3. Where the cause, or a contributory cause, of any damage is a persons reliance on a false or misleading statement or on any promise or advice, that damage shall not by virtue of the statement, promise or advice having been incorporated in a product be treated for the purposes of this Part as caused wholly or, as the case may be, partly by a defect in the product.

4. In subsection (3) above any reference to a statement or promise or to advice
   (a) shall include a reference to instructions for the doing of anything with or in relation to a product by an individual;

but

(a) in relation to a product which is or contains a device that makes or is used for the making of any measurement, calculation or reading, shall not include a reference to any measurement, calculation or reading made by or by means of that device; and nothing in that subsection shall prevent any statement, promise or advice from being considered in determining for the purposes of subsection (1) above what persons generally are entitled to expect in relation to any product to which the statement, promise or advice relates.

In any civil proceedings by this Part against any person ('person proceeded against') in respect of a defect in a product it shall be a defence for him to show:

(a) that the defect is attributable to compliance with any requirement imposed by or under any enactment or with any Community obligations; or
(b) that the person proceeded against did not at any time supply the product to another; or: that the following conditions are satisfied, that is to say – that the only supply of the product to another by the person proceeded against was otherwise than in the book of a business of that person's; and that subsection 2 (2) above does not apply to that person or applies to him by virtue only of things done otherwise than with a view to profit; or ...
(c) that the defect did not exist in the product at the relevant time; or
(d) that the state of scientific and technical knowledge at the relevant time was not such that a producer of products of the same description as the product in question might be expected to have discovered the defect if it had existed in his products while they were under his control; or
(e) that the following conditions are satisfied, that is to say –
(f) that the product was comprised in another product; and
(g) that the defect was wholly attributable to the design of the other product or to
(h) compliance by the producer of the product so comprised with instructions given by the producer of the other product.

5. (1) Any damage for which a person is liable under section 2 above shall be deemed to have been caused –
   (a) for the purposes of the *Fatal Accidents Act 1976* by that persons wrongful act, neglect or default; and for the purposes of section 3 of the Law Reform *(Miscellaneous Provisions) (Scotland) Act 1940* (contributions among joint wrongdoers), by that person's act or omission; or
   (b) for the purposes of the Damages *(Scotland) Act 1976* (rights of relatives of a deceased), by that person's act or omission.

Where any damage is caused partly by a defect in a product and partly by the fault of the person suffering the damage, the *Law Reform (Contributory Negligence) Act 1945* and section 5 of the Fatal *Accidents Act 1976* (contributory negligence) shall have effect as if the defect were the fault of every person liable by virtue of this Part for the damage caused by the defect. In subsection (3) above 'fault' has the same meaning as in the said Act of 1945.

SECTION 6: STATUTE AND COMMON LAW RELATED TO WEIGHTS AND MEASURES

Whilst the details of weights and measures legislation is an essential feature of quality management, it is more specialised than the topics that are elaborated on and the reader is advised to download the 1985 Weights and Measures Act and read it to familiarise themselves with the content. This will also provide a reference whenever required. There are two reasons for this. The legislation goes into a lot of detail on each of the legal measures of length, weight, volume, etc., which would essentially turn this section into a checklist. Much of it would have little relevance to some students but be very relevant to others. And also, it would make the section far too long bearing in mind everything else. The content is easy to read and to assimilate.

History

In previous centuries, most of the regulations related to weights and measures were for the protection of consumers and to assist in the regulation of trade. Any aid to industry in the achievement of traceability of measure was probably largely coincidental.

Hallmarking

The concept of hallmarking was introduced by an Act of Parliament in 1300. Under the hallmarking concept the measure of purity for gold articles is the carat. This is not a weight but a proportion equal to 1/24th. For example, 22 carat gold means 22 portions of gold to two portions of base metal.

There was also the problem of standardisation. At one time there were 23 different quantities for the measure of volume known as the 'gallon'. Unfortunately, the UK standardised on one of these and the USA another. Hence to this day, they represent different quantities with the USA version being some 20% smaller than that of the UK.

SECTION 7: OCCUPATIONAL SAFETY AND HEALTH (OSH)

Apologies but this is a long section, but it is most important that it is understood!

In many organisations, health and safety is seen to be a separate issue from quality, but we do not think so. How can you be a quality company if you do not consider your duty of care both to your employees and the community at large? Health and safety, environmental protection and quality are closely interrelated and can be

How is Health and Safety organised nationally?

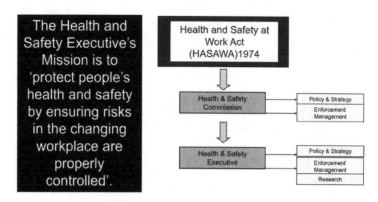

Figure 6.1   Health and safety organisation

referred to as 'integrated management systems'. Therefore, it is right that we should consider them here. In this section we will consider health and safety. Also, if you are using a Hoshin Kanri approach, then all of the issues such as environmental protection, quality and health and safety will be completely intermingled and not separable.

The job of the HSC is to protect everyone in Great Britain against risks to health or safety arising out of work activities; to conduct and sponsor research; promote training; provide an information and advisory service; and submit proposals for new or revised regulations and approved codes of practice

- HASAWA 1974 (an 'enabling act') absorbed existing laws into a coherent structure.
- Introduced the concept of safety representatives.
- Old 'regulations' were updated and new ones created, etc.

If the principles of Hoshin Kanri are followed, it can readily be appreciated that every KPI will have a safety-related element to consider and it is just not practical to separate the responsibilities for health and safety, environmental protection and quality as they are intrinsically inseparable. The law imposes a responsibility on the employer to ensure safety at work for all their employees. *Much of the law regarding safety in the work place can be found in the Health & Safety at Work Act 1974.* Employers have to take reasonable steps to ensure the health, safety and welfare of their employees at work. *Failure to do so could result in a criminal prosecution in the Magistrates Court or a Crown Court. Failure to ensure safe working practices could also lead to an employee suing for personal injury or, in some cases, the employer being prosecuted for corporate manslaughter.*

Control of Substances Hazardous to Health Regulations (COSHH)

COSHH requires employers to:

Figure 6.2   Impact of H&SE legislation

- Assess the risks to health from chemicals and decide what controls are needed
- Use those controls and make sure workers use them
- Make sure the controls are working properly
- Inform workers about the risks to their health
- Train workers
  What must the employer do?
  How can he/she achieve this?
  How can we recognise dangerous substances, or let people know that a product is hazardous?

The Noise at Work Regulations 1989 (Noise Regulations)

The Noise at Work Regulations 1989 (Noise Regulations) imposes a duty on employers to reduce risk of damage to hearing of employees from exposure to noise.

Action levels

- 'The first action level' means a daily personal noise exposure of 85 dB(A)
- 'The peak action level' means a level of peak sound pressure of 200 pascals
- 'The second action level' means a daily personal noise exposure of 90 dB
- Provision of information to employees
- Every employer shall, in respect of any premises under his control, provide each of his employees who is likely to be exposed to the first action level or above or to the peak action level or above with adequate information, instruction and training on:
  - (a) the risk of damage to that employee's hearing that such exposure may cause;
  - (b) what steps that employee can take to minimise that risk;
  - (c) the steps that that employee must take in order to obtain the personal ear protectors referred to in regulation 8(1); and
  - (d) that employee's obligations under these Regulations.

> ## Pause for thought
>
> Make a list of any activities in your work that might constitute a noise hazard that you think would have to be assessed.

Electricity at Work Regulations 1989

The regulations impose duties upon employers, self-employed persons, managers of mines and quarries and employees.

Adverse or hazardous environments

Electrical equipment that may reasonably foresee ably be exposed to:

(a) mechanical damage;
(b) the effects of the weather, natural hazards, temperature or pressure;
(c) the effects of wet, dirty, dusty or corrosive conditions; or
(d) any flammable or explosive substance, including dusts, vapours or gases, shall be of such construction or as necessary protected as to prevent, so far as is reasonably practicable, danger arising from such exposure.

Insulation, protection and placing of conductors

All conductors in a system that may give rise to danger shall either:

(a) be suitably covered with insulating material and as necessary protected so as to prevent, so far as is reasonably practicable, danger; or
(b) have such precautions taken in respect of them (including, where appropriate, their being suitably placed) as will prevent, so far as is reasonably practicable, danger.

Earthing or other suitable precautions

Precautions shall be taken, either by earthing or by other suitable means, to prevent danger arising when any conductor (other than a circuit conductor) that may reasonably foreseeably become charged as a result of either the use of a system, or a fault in a system, becomes so charged; and, for the purposes of ensuring compliance with this regulation, a conductor shall be regarded as earthed when it is connected to the general mass of earth by conductors of sufficient strength and current-carrying capability to discharge electrical energy to earth.

> ## Pause for thought
>
> Why not make a list of any activities in your work that might constitute an electrical power hazard that you think would have to be assessed?

The **Control of Asbestos at Work 2002 Regulations in October 2002** introduced the new duty to manage asbestos risk in non-domestic premises. This important legislation tackles the biggest occupational health killer in the UK – asbestos-related disease. Of the 3,500 people currently dying each year from such diseases, 25% have once worked in the building and maintenance trades and often would have worked unknowingly on or near to 'Asbestos Containing Materials' (ACMs).

Pressure Systems Safety Regulations 2000

An item of plant from the pressure system should be included in a written scheme of examination if its failure could unintentionally release pressure from the system and the resulting release of stored energy could cause injury. Each system is likely to be unique, but the following questions may help users to arrive at some decisions:

- Do the manufacturers of the plant or equipment forming the pressure system give guidance, instruction and the precautions to take for safe operation of the system?
- Could failure of any part of the pressure system cause someone in the vicinity to be injured by the release of pressure, fragments or steam?
- Does the pressure system contain any protective devices?

If the answer to any of these questions is 'yes', then those items of plant may need to be included in the written scheme of examination.

Management of Health and Safety at Work Regulations 1999 (Management Regulations)

This places an obligation on the employer to actively carry out a risk assessment of the work place and act accordingly. The assessment must be reviewed when necessary and recorded where there are five or more employees. It is intended to identify health and safety and fire risks.

Health and Safety (Display Screen Equipment) Regulations 1992

This applies to display screens where there is a 'user', that is, an employee who habitually uses display screen equipment as a significant part of normal work, or an 'operator', a self-employed person who habitually uses display screen equipment as a significant part of normal work.

Management of HASAW Regulations 1999

Nothing in these regulations shall apply to or in relation to:

(a) drivers' cabs or control cabs for vehicles or machinery;
(b) display screen equipment on board a means of transport;
(c) display screen equipment mainly intended for public operation;
(d) portable systems not in prolonged use;

(e) calculators, cash registers or any equipment having a small data or measurement display required for direct use of the equipment; or

(f) window typewriters.

Pause for thought

Identify examples in your place of work where you think that these regulations will apply.

Manual Handling Operations Regulations 1992 [MHOR]

This require employers to:

1. Avoid hazardous manual handling operations so far as is reasonably practicable.
2. Make a suitable and sufficient risk assessment of any hazardous manual handling operations that cannot be avoided.
3. Reduce the risk of injury from those 'unavoidable' operations (referred to in 2 above) so far as is reasonably practicable.
4. Provide employees with details concerning the loads they are to handle.
5. Review the assessment to keep it up to date.

The facts

*   More than one-third of all over three-day injuries reported to the HSE arise from manual handling
*   Many manual handling injuries build up over a period rather than being caused by a single handling incident i.e. they are cumulative.

Employers' responsibilities

The law

The Manual Handling Operations Regulations 1992 came into force on 1 January 1993. *They require an employer to undertake a suitable and sufficient assessment if there is a possibility of risk from the manual handling of loads.*

AVOID: Need for hazardous manual handling as far as is reasonably practicable.
ASSESS: Risk from any hazardous manual handling that cannot be avoided.
REDUCE: Risk of injury from any hazardous manual handling as far as is reasonably practicable.

Pause for thought

Identify the lifting operations carried out in your company, (or your department if more appropriate) and the equipment used.

Personal Protective Equipment at Work Regulations 1992

The effect of the PPE at Work Regulations is to ensure that certain basic duties governing the provision and use of PPE apply to all situations where PPE is required. The regulations follow sound principles for the effective and economical use of PPE, which all employers should follow.

Work place (Health, Safety and Welfare) Regulations 1992

These regulations deal with any modification, extension or conversion of an existing workplace. The requirements include control of temperature, lighting, ventilation, cleanliness, room dimensions, etc. The regulations also provide that non-smokers should be allocated separate rest areas from smokers.

The Provision and Use of Work Equipment Regulations 1998 (PUWER)

These regulations deal with minimum standards for the use of machines and equipment about suitability, maintenance and inspection. The regulations will also cover mobile work equipment from December 2002. Generally, any equipment that is used by an employee at work is covered, for example hammers, knives, ladders, drilling machines, power presses, circular saws, photocopiers, lifting equipment (including lifts), dumper trucks and motor vehicles. Similarly, if you allow employees to provide their own equipment, it too will be covered by PUWER and you will need to make sure it complies.

Pause for thought

Identify key items of equipment relevant to these regulations in your work.

Lifting Operations and Lifting Equipment Regulations 1998 (LOLER)

The regulations aim to reduce risks to people's health and safety from lifting equipment provided for use at work.

In addition to the requirements of LOLER, lifting equipment is also subject to the requirements of the Provision and Use of Work Equipment Regulations 1998 (PUWER), Lifting Operations and Lifting Equipment Regulations 1998 (LOLER). Generally, the regulations require that lifting equipment provided for use at work is:

- Strong and stable enough for the use and marked to indicate safe working loads
- Positioned and installed to minimise any risks
- Used safely, i.e. the work is planned, organised and performed by competent people
- Subject to ongoing thorough examination and, where appropriate, inspection by competent people.

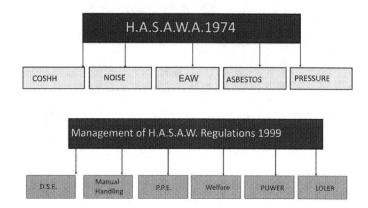

Figure 6.3   HASAWA structure

Pause for thought

You might consider making a list of the main H&SE regulations that should be highlighted in your developing quality strategy and if you are developing a Hoshin Kanri approach, then what impact might these regulations have on the priorities of your KPIs?

SECTION 8: PRODUCT CERTIFICATION SCHEMES

CE marking

Many products such as machinery, toys and medical devices must meet legal requirements before they can be sold within the European Community, and must carry CE marking. CE marking attached to a product is a manufacturer's claim (self-declaration) that it meets the European legislation for health, safety and environmental requirements. Some products carry both a national and CE marking. This indicates that a national body has independently tested them against the appropriate standard. The CE mark is not a European safety or quality mark.

The BSI Kitemark

Within the United Kingdom the most famous product safety marking is the Kitemark. The Kitemark is only applicable to those products that have been tested by BSI. There are other accredited testing organisations on the UKAS website (http://www.ukas. com/about-accreditation/accredited-bodies/).

For specific information regarding the Kitemark and its uses see:

www.bsigroup.com/en-GB/kitemark/

There are other relevant standards (there are many others and some derivatives of these but these are the main core standards).

SECTION 9: RISK ASSESSMENT

Risk assessment is a process where hazards are identified, and risks evaluated, with the objective of eliminating or reducing the risks as low as is reasonably practicable

Figure 6.4   Managing risk

Pause for thought

Based on the foregoing, Check to see if you can locate any serious risks in your organisation that might not be being properly managed
Score the hazard using a rating system. Here is an example:

PROBABILITY 1 Almost impossible (<5% change)
             2 Unlikely (5–25% chance)
             3 Likely (25–70% chance)
             4 Probable (70–95% chance)
             5 Inevitable (>95% chance)
FREQUENCY   1 Arises infrequently (annually or less)
             2 Present monthly
             3 Present on a weekly basis
             4 Present daily
             5 Permanently present
SEVERITY   1 No injury
             2 Minor injury with less than 3 days lost time
             3 Minor injury with more than 3 days lost time (report)
             4 Major injury (report)
             5 Death or major injury with permanent disablement (report)

Rating

Number of people   Weighting factor
exposed
1–4   Risk Index x 2

5–9	Risk Index x 3
10–24	Risk Index x 5
25 or more	Risk Index x 10

Example – action levels

Risk index	Action level	Action required
90 or more	1	Immediate action to reduce the risk
70–90	2	Action within 7 days
30–70	3	Action within 3 months
Less than 30	4	No action required

This concludes the section on legal aspects but, before moving on, it is important to reflect that whilst we have looked at some of the critical laws and regulations here, the full list of relevant legislation is not only almost limitless but is also changing all the time. Regardless of all the laws and regulations that apply in your respective markets, do not overlook the fact that you are constantly vulnerable to Common Law actions at any time and the 'duty of care' and negligence accusations are constantly present. Just because you do not think you have done nothing wrong, does not mean that others might have a different opinion however unfair you might believe that to be. In the end, it might be the court that decides this.

It is a topic that those with any interest in their company's reputation regarding the opinions of society at large must be constantly vigilant about. Anyone could find themselves having to defend both their personal and their organisation's reputation at any time. The risk of a catastrophically expensive product recall is always there and can hit you in your sleep. Take the situation with Celotex. This was a company making building insulation materials, an excellent product with a huge future given the ever-increasing market for environmentally friendly ecosystem products. Suddenly, they were caught up in the Grenfell Tower disaster inquiry. Look at the Trading Standards Office website on product recalls on any day and scroll down the number of household names you will find. Never mind whether your company has ISO 9000 certification, it will probably not be mentioned in any inquiry into an incident, nor will it protect you from the likelihood of such an event. Probably, all the companies listed in the product recalls are certified.

Summary of learning

In this chapter you have learned the legal definition of a defect in statute law and its relevance to the design, manufacture and distribution of goods and services.

- Legal framework – UK, EU

- The principles of common and statute law
- The role of precedent and case law in the law of tort
- Common law applied to quality. The common law relationship between buyer, seller and third parties
- Statutory law and its impact on the responsibilities of an organisation and quality professionals
- Common statutory requirements
- Sale of Goods legislation
- Consumer protection legislation
- EU directive on product liability
- Occupational safety and health
- Risk assessment.

7 *New product development*

LEARNING OBJECTIVES

The learning objectives are to be able to design and coordinate quality management activities that support the product or service through its whole lifecycle from conception to disposal, supporting the various business functions involved in each part of the process.

The learner will:

- Understand the stages in the service/product lifecycle and their role in the achievement of competitive quality products and services
- Understand the quality-related concepts involved in the new product development process
- Understand make or buy decision-making criteria
- Recognise potential costs due to poor design and development processes.
- Understand sequential and concurrent design
- Selection and use of quality tools throughout the product lifecycle.

SECTION 1: THE ECONOMICS OF THE NEW PRODUCT DESIGN PROCESS

It begins with the evaluation of whatever market research information may be available and runs through all of the proposed activities of conceptual design, functional design, design for operations, the 'make or buy' decision and the relevant decision criteria, vendor evaluation and control, internal operations, packaging, shipping and delivery, after sales product support and, eventually, through to final disposal at termination of its useful life. This can be a very long process in some cases and may even go beyond the expected working lifetime of many of those involved, for

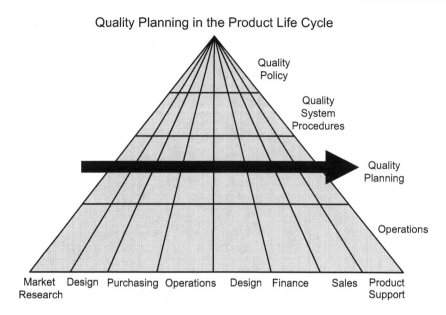

Figure 7.1   Quality planning

example, long life and long disposal periods as in the case of nuclear power stations. At the other extreme, the entire process may be only a matter of hours, days or weeks. For example, the planning of newspapers and other media forms.

The cost of design failures

It is a fact that approximately 80% of the cost of poor quality has its roots in the design stage. Typically these may be due to lack of information related to customers' needs, very commonly skimping on product testing in order to reduce the time to market for new products, or wrongly selecting inappropriate manufacturing processes. Chapter 1, refers to the British Trading Standards list of product recalls that were topical at the time of writing this book. Many of these recalls are attributable to problems that originated in the design stage and this is avoidable. In this chapter we will concentrate on how through good practice not only can these costs be minimised but, in many cases, the time taken to bring a new product to market can be minimised.

Case example: a company that manufactures sophisticated electro mechanical medical equipment used QFD (quality function deployment), which is explained in this chapter, to replace an existing design due to falling sales. Apparently, it took five years to bring that product to market from the initiation of the design process and only after a long string of design change requests from production and further downstream. Even then, there were numerous further design changes caused by failures in the field resulting in a series of product upgrades, each of which resulted in the need for a huge spares inventory. Using QFD to design the successor, the new product, which was even more sophisticated, took just one year for the complete design process and product launch. There were virtually no design change requests and two years after launch no field mortality failures requiring an upgrade.

If poor quality has been built in at the design stage then no amount of care or inspection will take it out later, therefore that aspect of quality planning is critical to success. Between each of the key stages in the design process, it is necessary to thoroughly review all quality critical features before progressing to the next stage. This is an iterative process because if some things are not quite right, they must be readdressed until they are.

This is where tensions frequently arise between the quality department, design and sales. Sales are anxious to get an innovative new product into the market place to obtain the benefits of prior franchise. This always puts intolerable pressure on the design department. But designers are also enthusiastic about their new concept and are also anxious to see it in use as soon as possible. It is not difficult to see how the temptation to jump stages is difficult to resist.

It is the quality function's responsibility to restrain both these risky temptations ~~drives~~ until it is sure that the product or service is right. This often requires imaginative and often exhaustive and time-consuming testing. If not, the consequences are sometimes disastrous, for example, damaging recalls, plane crashes and other tragic disasters.

The new product design cost model

Note that the data in Figure 7.2 was based on a study made by the Ford Motor Company in the 1980s comparing the typical costs associated with new product development in the West compared with Japanese Companies. In general, Western companies invest less at the design stage than their Japanese counterparts but once the product leaves design and moves beyond product release into the market, the

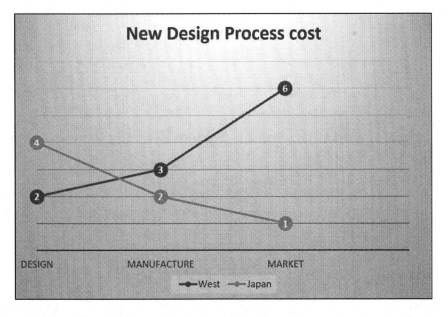

Figure 7.2   New design costs

costs to the Japanese companies drops dramatically whilst the Western equivalents increase significantly. This has probably improved considerably in recent years in the automotive field, but this is possibly not the case with other forms of manufacture.

The shape of the curve for Western designs is predictable. It can be seen that the costs increase the further downstream the defects are located from their source. If a design deficiency is identified during the design stage, it can easily be corrected with a few strokes of the mouse or a pencil and rubber. If it is not detected until operations, then, first of all, there are delays that can sometimes be significant plus the inevitable 'whose fault was it'. Then there is the possibility of creating new jigs, fixtures, possible discussions with suppliers and obsolete stock. If it is not discovered until it is in the market, then the potential costs are almost limitless depending upon the situation. Re-read Chapter 6, especially on 'the duty of care' and product liability. It makes sense to be very diligent at this stage – everyone's future depends upon it, lives sometimes.

SECTION 2: NEW PRODUCT DEVELOPMENT PROCESS PLANNING

Normally, quality planning begins with the creation of a flow chart to determine the key points where there are likely to be quality-related issues. These flow charts may sometimes be quite complex or simple depending upon circumstances. For example, the flow chart in Figure 7.3 appears somewhat complex, but they can be much more complex than this for large operations for example, the design, manufacture and assembly of a nuclear submarine. The quality plan for such products is hugely complex. However, complex or not, all of them including the simple ones all follow the same basic principles. Figure 7.3 shows the product development process developed by DHI for the creation of new courses.

The initial steps will vary, depending upon whether the concept is determined by the customer or the manufacturer/service provider. Subsequent stages will usually be dependent upon the complexity of the product/service and the degree of innovation. The activity of quality planning usually begins at the first design-review activity. The procedure of design review applies to:

(a) Feasibility studies for new products/services and new product/service designs at various steps in the design process.
(b) Existing products/services at intervals after their initial circulation.

The objective of the review is to bring together specialists from all relevant functions and, when appropriate, the user, with a view to optimising the design in terms of satisfaction of customer requirements, reliability, ease of manufacture, ease of assembly, ease of maintenance, service, appearance and safety.

Basically, there should be a minimum of three design-review activities in the design process. These are usually iterative, so each might be repeated several times until it is finally signed off as being satisfactory. Sometimes these reviews are referred to as 'gateways'. In this case, the design cannot progress to the next stage until it has passed all the criteria identified in the previous stage.

SYSTEM **PROCESS** **PRODUCT**

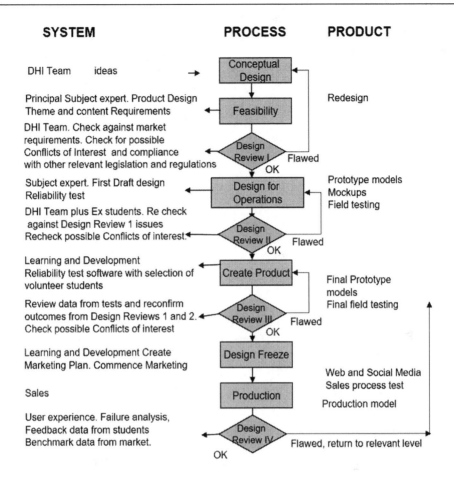

Figure 7.3   Product development process

The feasibility stage is the first, when all existing knowledge of customer requirements is compared with the feasible methods of satisfying those requirements. This is followed by some intermediate reviews, of which there may be several, in which the results of feasibility studies, prototype-tests information, performance claims, design life and reliability data are evaluated.

The final review will determine whether the completed product conforms sufficiently closely with the requirements of the customer. This review is also concerned with the manufacturing process and 'make or buy' decisions. Ideally, this requires a database of process capability data from existing processes, decisions as to whether capital investment is needed or to avoid it by outsourcing. Supplier evaluation and supplier data bases are interrogated to select the optimum suppliers. Considerations include methods, materials processes and assembly methods in order to ensure the optimum cost and quality of manufacture.

The specialists in a design-review team should include at least one of each of the following: designers/product developers, quality engineers, production planners, production engineers, specification and standards engineers, purchasing officers, safety officers, technical sales representatives, market research, cost accountants, after-sales service personnel.

The following checklists show suggested items that might appear at each of the three stages of design review. Obviously, this could be extended or reduced according to circumstances.

Design review 1

1. Interpretation of the potential customer's requirements in terms of performance, expected life and reliability.
2. Cost and selling price limitations.
3. Details and consideration of the state of the art, or technological limitations of all possible concepts for meeting the customers' requirements.
4. Expected life of the product before technological obsolescence.
5. Consideration of all constraints, including environmental, safety legislation and considerations, assembly, operating, storage, transportation, size and interchangeability.
6. Development programme, including research and testing of new concepts.
7. Relation with mating products and aesthetic considerations.
8. All information related to competitor's offerings, strategy and also potential competitors in the case of completely innovative products or services.

Intermediate design review (design review 2)

For a complex product, this stage might necessitate the dissemination of a considerable amount of data. Should this be the case, the product should be considered item by item, or by individual assemblies, by specialist groups brought together for the purpose. This will save wasting the time of those only concerned with other aspects of the product and will allow development to continue where possible.

The following is included for illustrative purposes simply to demonstrate the detail that is required.

The product characteristics should be determined during these intermediate reviews. Essential considerations at this stage should include the following:

- Failure mode and effect analysis (FMEA), including reference to dangers that might arise in respect to product liability and Health and Safety at Work Act (HASAWA) risks. In the food industry FMEA is known as HACCP (hazard analysis and critical control point) for food safety.
- Consideration of data relating to product degradation and contamination, including fumes, biological attacks, moisture, and health hazards, not forgetting pollution.
- Analysis of prototype test data, including life tests, accelerated life tests, and environmental tests.
- Determination of performance parameters of items, assemblies and the complete product.
- Value engineering considerations of components, and cost reduction possibilities based upon the following questions.

(a) Does its use contribute value to the product?
(b) Is its cost proportionate to its usefulness?
(c) Does it need all its parts?
(d) Is there anything better for the intended use?
(e) Can a usable part be produced at lower cost?
(f) Can a standard item be used?
(g) Is the tooling adequate, considering the quantity used?
(h) Can another dependable supplier provide it for lower cost?
(i) Is anyone buying it for less?
(j) Does it increase variety unnecessarily?
(k) Could another item from the standard range fulfil the function sufficiently well?
(l) Has standardisation been adequately considered in the range of component parts and dimensions?
(m) Is it interchangeable with other designs?
(n) Could alterations in design obviate the need for special tooling?
(o) Does the design facilitate ease of assembly and ease of maintenance?
(p) Has tolerance analysis been carried out and are tolerances compatible with both design requirements and manufacturing compatibility?
(q) Has it been designed in such a way as to prevent the possibility of its being assembled the wrong way around, or in the wrong product?

Review of manufacturing methods

The objective of design review of manufacturing methods should be to:

(a) Reduce the number of parts (simplification).
(b) Reduce the number of operations and length of travel in manufacturing.
(c) Utilize the best available materials.
(d) Use of least costly processes.
(e) Analyse tolerance and ensure use of statistical tolerancing where appropriate.

A checklist of factors to determine minimum cost designs should be run through at the appropriate design review. Check that all drawings, specifications, test reports, etc., are clearly identifiable, and that issue numbers of documents and their circulation is strictly controlled.

Final design review (design review 3)

This review will include a complete re-approval of all the items listed in the previous stage, together with a detailed consideration of all manufacturing requirements. Comprehensive data obtained from tests on pre-production models should be analysed. In the case of manufactured products, it is essential that these tests are carried out on products manufactured by tooling and production personnel rather than those previously produced in the model shop by craftsmen. **NOTE! In the rush to put a new product line on the market, this stage is frequently overlooked.**

The risks inherent in this practice are considerable. Craftsmen may easily alter a dimension in order to ensure a good fit, and this may not always be written into the production drawing. An internationally known motor-car manufacturer recently suffered the consequences of such an occurrence. The result was a catastrophic seizure of the components in the gearbox, which resulted in three fatal accidents, and an extremely expensive product recall.

Items to be considered in addition to the above must include:

- A review of all long-delivery items, such as special materials and special tooling
- Consideration of purchasing needs, including vendor appraisal of new suppliers
- Critical review of all tolerances and other specification limits
- Consideration of all inspection and test requirements. The appropriate visual standards must be set and agreed, and the appropriate means of assessment determined.

Items to be considered in addition to the above must include:

- Analysis of training requirements for all production quality control and inspection personnel
- Review of the quality plan for the manufacture of the product
- Review of the packaging and labelling arrangements.

The following questions should be asked at this point:

- Are the potential dangers that may be inherent in the use of the product clearly marked indelibly on the product, and on its carton or container, and on the instruction leaflet?

The importance of this aspect is well illustrated by a legal case in 1971, Vacwell Engineering Ltd v. BDH Chemicals Ltd. The defendants manufactured a chemical that they marketed for industrial use in glass ampoules bearing a warning label 'Harmful Vapour'. Following discussion with the defendants, the plaintiffs used the chemical in their business of manufacturing transistor materials. To prepare the chemical, the labels were washed off the ampoules in sinks containing water and a detergent.

During this procedure there was an explosion, resulting in the death of the operative and extensive damage to the plaintiff's premises. It appears that one of the ampoules had been dropped into the sink where it had shattered, releasing the chemical into the water. The ensuing reaction had broken the glass of the other ampoules, which had then also mixed dishwater and caused the major explosion.

At the time of manufacture the effect of water on the chemical was not known to the defendants, but the dangers had been detailed in scientific lectures dating from the previous century. It was held that: defendant manufacturers were liable in respect of the damage suffered, because they had failed to carry out adequate and proper research into the scientific literature and to give full warning of the changes accompanying the use of the product. Research consistent with the exercising of 'reasonable care' would have reduced the likelihood of the being mixed with water.

Another example of this is contained in an American case involving a two-gallon container containing two gallons of carbon tetrachloride carpet cleaner. Sold under the brand: name 'Safety Clean', it had on its packaging the word 'Safety' so clearly marked on all four sides, that the word 'Caution', the admonition against inhaling fumes, and the instruction to use only in well-ventilated places, seemed of comparatively minor import and were missed by the consumer, and resulted in her death. The manufacturer was aware of the danger but misapplied his intention in not carefully expressing his wishes and warning to the consumer.

Are the packaging instructions sufficient to ensure that the product will not be exposed to the elements sufficiently or degrade, in any way, during storage or transit, and will the product be protected against all possible forms of contamination?

Are the labels sufficiently distinctive, such that the product is unlikely to be confused with another product? This is particularly important with chemicals and pharmaceutical products. If containers of substantial construction are used by the company's customer to subsequently pack that customer's own products, it must state clearly on the container that they must not be used again.

It is possible that an ingredient in the packaging material may react with the product, and it is not certain that the customer could not sue the supplier of the box on the basis that it was purchased with the supplier's goods. It is also possible that the containers may collapse when stacked if used subsequently. Again, a product-liability or a Health and safety at Work Act (HASAWA) problem could arise. The original supplier can only safely protect himself by stating clearly on the container that it may not be used again, and either ask for its return, or, better still, collect it from the customer after delivery.

Pause for thought

Some organisations have very sophisticated new product design processes involving what are referred to as 'gateways'. These are also known as 'design review' activities where the decision is made either to refer the product back to repeat or correct an earlier activity or to pass it to the next stage. Where we have shown 'design review', the term 'gateway' could be substituted. Many of these organisations have something like seven or so gateways in which there is considerable scrutiny of the design. Check your own NPD process and if you do not have such a structure, try to create one based on the flow diagram that we demonstrated earlier. Create a version of your own and create a quality critical checklist for each design review stage.

SECTION 3: STAGES IN THE DESIGN PROCESS

Invariably, no matter how meticulous we may have been at the planning stage, the probability will be that nothing is perfect. In fact, in many cases we will be very far from perfect. Basically, there are three key phases in the design process and problems can originate in each.

The three key stages in the design process

- Conceptual design
- Functional design
- Design for operations.

Conceptual design – typical problems

1. Relationship with market research. Problems here are on the interface between the two functions. Professional market researchers learned their skills in an education system that taught them all about social classes, such as A, B, C1s and C2s, etc. and how to determine their needs and to use non-parametric statistics to prove their arguments. These non- parametric tests have a very unusual vocabulary not generally known and probably not by many conceptual designers. So, their reports are full of data referring to the use of such specialised techniques as Tukey tests, Kolmogorov–Smirnov tests, Wilcoxon signed-rank tests, etc. Conceptual designers generally went through a very different educational process. Probably, they learned none of these techniques and therefore the market researchers reports probably appeared like a foreign language. However, the conceptual designer was, in any case, probably very mistrustful of the market researcher. After all, he would more than likely see himself as being the expert in this field. He thinks he 'knows' what the customer needs regardless of what the customer thinks he wants or not. So, our designer designs a perfect product, which sometimes almost perfectly misses the requirements of the customer! A quality problem right at the beginning of the process! Technically it may be excellent, but the customer may remain unimpressed. For example, in the mid-1960s, the British Motor Corporation was rapidly losing market share to Nissan or Datsun as it called itself in those days. They could not understand why. Their equivalent to the Datsun family saloon was far superior technically. After all, it had 'hydrolastic suspension'! The Datsun was very basic, 1930s technology it was claimed by its distractors, but it did have a radio on the dashboard. The typical customer knew nothing about the sophistication of the suspension system, but they did recognise a radio when they saw one! So, they bought the Datsun! BMC never did get to understand this lesson and it was not long before they faded from the market altogether.

2. Relationship with sales. This problem is often far more serious. The sales department know that they can sell anything if it is exciting new and different. Where might there be something exciting, new and different? Conceptual design, of course. So, the super salesman, polka dot tie, moustache, hankie in breast pocket and dressed to kill, surprisingly makes friends with the conceptual designer. Strange relationship. One is extremely extrovert and the other probably quite introvert. Not a typical match. So, the super salesman gleans all the embryonic plans form the designer who is proud of what he is doing and loves to share it. Before you can blink, the salesman has promised the potential customer that he can have this product in only months, even weeks possibly. He does not feed this information back to the designer but to top management as it is they who pay his bonus!

The next the designer knows is that he is under enormous pressure to get the design completed in the minimum time. Cut corners, just get it finished.

Functional design

The job of this function is to verify that the product will do what it is supposed to do, not just on day one after it has been put to use but continuously and reliably throughout its intended design life and then through to the means of eventual disposal. In recent times, society has become increasingly concerned with disposal of worn out or obsolete products in an environmentally friendly way. The most desirable situation is to make the no longer wanted product recyclable. In the case of physical products this usually involves several exhaustive tests, usually to destruction. The failure mode is noted and the possible consequences. Ideally, the tests should also include what happens when there is improper but foreseeable misuse of the product and its consequences. This is partly why Chapter 6 on the legal aspects was included. In almost all disasters involving product failures, the focus eventually centres on whether this function was properly carried out. However, this function more than any is under enormous pressure. It is where the damage done by the false promises from our super salesman materialises. Questions put to the staff include 'how long will this test take'? If it is a destructive test, it could take months, or even years sometimes. In the latter case there are ways around that as will be explained later but, frequently, the pressure is such that the test is skipped altogether or done very badly with the consequence that the product has a potential disaster already built in waiting to happen. Product recall list abound with these examples as do the News Headlines following a tragedy of some kind almost daily somewhere. At the end of this chapter, some of the simpler ways of reducing these risks will be explained.

Design for operations

When the design has been through the first two stages, the next phase is design for manufacture. The first question is make or buy? Do we have the capability to make it ourselves? If not, should we invest or find a supplier who has the expertise or who might be willing to invest in creating the processing resources. Before all of that however, in many cases, the design features should be studied by those with operations experience to see if there are any unnecessary features that increase cost but are of questionable value. Typically, this often includes the allowable variation on the dimensions of the product. For example, maybe a dimension is nominally 50 mm. Of course, it is impossible to achieve exactly 50 mm every time we try. Depending on the precision of the manufacturing process, the accuracy of the gauges and the skill of the workers, there will be some variation from piece to piece. What the operating forces need to know is how much variation is acceptable. Frequently, the designer is not entirely sure. However, he does know that the more accurate it is the less likely there are to be problems with interfaces with other parts, etc. So, the temptation is to put very tight tolerances on the dimensions. Unfortunately, the tighter these tolerances are, the more expensive will be the product to produce. If they are very tight, it might require investment in more sophisticated machinery, gauges and more highly skilled operating forces, not to mention the additional scrap or rework if the parts are outside specification. So, one of the tasks at this stage is to challenge manufacturability. If this is not done adequately, and the design goes into operations, there will almost always be howls of protest 'who designed this then'? 'They want to come down here and try to make it.' So, the drawing change requests begin to flow, delays abound, the delivery dates get pushed ever farther back and customers get more and more angry or may cancel altogether.

Operations. Well, problems in this part of the process are more well documented in quality literature than at any other stage. There is no doubt that there are huge opportunities especially where large volume production is happening. This is the area where MUDA is talked about and costs such as long set-up times, so-called work in progress (this is a marketing term for work that is not progressing otherwise it would not be there!), incoming goods stock and finished goods stock. It would be good to cover this subject in this book but it is covered in one of my other books, *Just in Time*, printed in some other languages and published by this same publishing house.

Pause for thought

Look into your operations now if you are a producer. What percentage of your plant and equipment is producing added value product at this instant? Use a technique called activity sampling to do this.

How many pieces of equipment are in the process of being reset? How long, on average, do resets take? If there is a lot of work in progress at some locations, are there any bottlenecks? What is the total value of incoming goods stock and finished goods stock? How many stock turns are there per year? Probably, when you have done that you will have found many opportunities to use the improvement process in Chapter 4 regardless of labels such as Six Sigma and lean. You have probably walked though those departments many times, but these items have never jumped out at you. Also, when you have time, go to the design department, check out the files on drawing change requests, check past designs and check the change control history or minutes of design review meetings. What does it tell you?

Post delivery

Now is the time to check customer complaint and customer return information. This can lead you to the gold in the mine. So, we find that we designed a perfect product, manufactured a perfect product but failed because we overlooked the importance of teaching our customer how to use it! How many times have we been there when one of our children opens a much sought-after Christmas or birthday present, asks you to assemble it or set it up but the designer who had an IQ of 150 assumed that this was true for every user too and forgot to teach you how to get best use out of it? Actually, this is not just children's presents. It is most things we buy these days. Either the print is too small to read, or we are told to download the manual from an obscure website, but it happens we are out of range where we are or that we get a manual two centimetres thick only to find that there is only a single page in our language and the rest of the book is in every language under the sun. If it is something that we must assemble, and we get it wrong due to unintelligible assembly instructions, is it our fault or whose is it? It is a quality issue, that is for sure. One other thing is also for sure. If anything is skimped in any of the three phases of design, then as soon as operations commence problems will abound. Some will be very serious and obvious. Others may be just as serious or even more so but not at all obvious until a lot later.

Pause for thought

It is likely that from the previous text there will be a lot for you to record. Where is the notebook? When you find something, think, which previous chapter in the book is also relevant? Also, remember that there is still Chapter 8 ahead, which covers the supply chain. There is work to do!

SECTION 4: QUALITY FUNCTION DEPLOYMENT (QFD)

When carried our efficiently and correctly, quality planning in the product lifecycle makes use of a concept known as 'quality function deployment' (QFD). If you are not familiar with it already, it might look complicated, but it is not and well worth the time it takes to become familiar with it. You might get a group of colleagues together and work through the rest of this chapter together. Before attempting to use the tools, run a session in which the participants are required to consider the following observations to create the right mindset.

The QFD concept differs from the tools of continuous improvement in as much that the latter is concerned with the improvement of something that already exists whereas QFD is the systematic and sequential use of a collection of tools which can be used for planning something new.

Notice that the branch on the left in Figure 7.4 will use the improvement process covered in Chapter 4 whereas the branch on the right for new products will follow the branch on the right. This might be a new physical product, but it could also be the development of anything, even a local fete, a barn dance, anything where you are virtually starting from a blank sheet of paper, or, as was explained in Chapter 5, the creation, development and maintenance of Hoshin Kanri in an organisation. Indeed, several of the tools of QFD have already been used by us to create our programme of courses starting with our vision statement and the conversion of big ideas to specific 'key performance indicators' (KPIs).

QFD is a term used to describe the systematic application of a series of techniques for ensuring quality throughout each stage of the product or service development process, starting with market research, feedback on customer satisfaction and ultimate product disposal. In other words, it is a method for transforming market demands

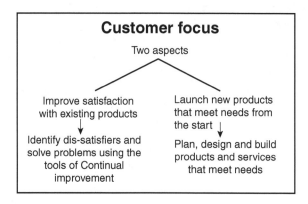

Figure 7.4   Customer focus

into design objectives and measurable quality goals for each stage in the product/service development process.

QFD is a process that begins with attempting to identify the vague needs of the ultimate target beneficiary of the intended product or activity, crystallising these into specific tangible and measurable goals and through a synthesising process progressively organising these needs into specific and measurable actions and responsibilities and a plan for implementation.

Pre-QFD tools session

Step 1

First, it is advisable to attempt to gather a team including a representative of all those directly involved personnel and, if possible, a representative potential customer to decide what you are intending to achieve and who is the intended customer. For example, it could be to create a new product, to implement Hoshin Kanri; to organise a conference; to design and implement totally new pay structure for the organisation or the development of a completely new product or a garden party, the applications are almost unlimited. We have used the development of a radio-controlled racing car as an example in the text below.

Step 2

Attempt to determine customers (or impacted party's) stated and likely unstated needs. At this stage we should consider the questions in Figure 7.5 to enable us to take a broad view of the situation.

External key issues

We tend to think of the direct purchaser of our products or services when we use the word customer. But we need to take a much broader view and consider everyone who

Playing to win

Supply Chain Management

Do we involve key suppliers in our customer focus policy?

Which suppliers should be involved?

(suppliers of services and know-how, as well as products)

Customer

Do we know what our customer <u>really</u> wants?

Do key employees know what <u>their</u> customers <u>and</u> the ultimate customers really want?

Do we have a POLICY for employee involvement?

Competitor

Do we know who our <u>real</u> competitors are?

Do key employees know who the competitors are?

Do we know competitors strategies for increasing market share?

Do we know what they are doing to make our life difficult?

Figure 7.5   Supply chain questions

may be impacted by our intentions since their needs or possible objections may put a number of constraints on what we plan to achieve. We need to identify those now!

These additional, but important, groups we may call *'stakeholders'* and include our own workforce, for example, government regulations, non-exclusivity policies (they may multi-source), their own stakeholders policies, local environmental groups, the local community (a proposal for a new airport terminal is a case in point), the opinions of the market and our customers' own customers, etc. The company has customers but so do people at each stage in the operations.

The next downstream operation is the customer of the one before it. These are the 'internal' customers. The same questions apply.

- Who are **our** customers?
- What are **their** needs?
- Do **we** satisfy their needs?
- How do **we** know?
- Have **we** ever asked?
- How much of the difference is determined by **our** suppliers?
- Do **they** know?

A lot of questions – do we have the answers? How might we get them?

The customer related situation today:

- New products are expected to be better, faster than previous models.
- We expect better service and technical support.
- We do not expect products to fail but when they do, we expect the supplier/producer to take care of us.
- Our expectations are continually moving upward.
- Business must continually improve at an equal or greater rate than customer expectations.

We can drive customer expectations! John D. Rockefeller is reputed to have given the Chinese a million oil lamps free, the reason being that he wanted to sell them oil! There are situations where we can induce our potential customer to want our product!

Perceived quality

It is the customer's 'perception' of our performance that is important, not our actual performance. If the customer perceives that we are poor when we are good there is no point in making improvements. We need to change the perception!

Example

Reels of steel leave a steel plant based in the desert, unwrapped and tied to the platform of open-backed trucks. They are driven across the desert for about 100 miles with no protection and are delivered to their customer, an automobile company. This led to a perception that the steel can be rusty even though the climate was so dry that the steel did not need protection. The reason for this perception was because they saw foreign steel – presumably because it travelled by sea – protected from the salty air on board the ship by wrapping it in oily paper.

We knew that the steel was perfectly good, but if that is the customer's perception, then making the steel shinier will not make any difference. We must change the perception, not *necessarily* the quality, which may be OK anyway! Therefore, the steel producer did nothing to the steel but wrapped it, as was the case for the foreign steel. There was an immediate reaction from the client that the steel was now of significantly improved quality, but it was exactly as it had always been!

Pause for thought!

Do we know what our customers perceptions of our products or services are? We cannot possibly guess other people's perceptions. We have to find a way to find out!

This is not always easy, and it can be expensive but not as expensive as getting it wrong, which might cost us our business!

What do our customers want?

There may be a difference between the customer's stated needs and their **real** needs. Do we know what they are? For example, the customer may state that they want to open a bank account. The **real** need will be to be able to transfer money easily or to arrange a loan, buy a car or other requirements – to have easy access. If 'online' banking is used, the customer wants the software to be intuitive, to be listened to. If we have problems, we want the bank to call us back promptly, to feel they actually care and that we are not just a statistic whether we invest billions or single figure amounts.

What are the real needs? The bank that gets closest to the real answer will increase market share and customer loyalty. These needs can change from time to time so we must continually review. For example, the cheque book is less important now that we have electronic transfer. What changes does this mean for skills, staffing and logistics, etc.? The bank that gets it wrong may go out of business! Unfortunately, it sometimes feels as if they have all ganged up to offer the same unsatisfactory service. This goes for airlines as well.

Who is the customer?

The customer will be a cast of characters, for example, a hospital. On the face of it, this is one customer, but it has many faces: the buyer or purchasing department, laboratories, maintenance and repair, training, doctors, surgeons, nurses, cleaning staff and the patient, etc. Each will have their own needs and may deal with different departments and have their own impressions of our organisation but often for very different reasons. Do we know what they are? Have we ever tried to find out? How?

What do they want?

The buyer: he will be interested in price, offers, delivery schedules, invoicing and other relevant documentation. He will deal directly with the sales department of our company and it will be their behaviour that colours his view of the organisation.

Supply quality assurance: they will deal directly with their quality counterparts at our organisation and their view will be coloured by the quality system as they see it and the way they are treated.

Quality assurance may be good but if they are badly treated by security, staff arrangements, etc., they may form a poor opinion. It is a fact that poor quality is remembered long after the price has been forgotten.

The role of the customer

If we do not have a customer, we do not have a business. This is a seemingly obvious comment on the face of it, but the fact remains that even in a highly competitive market there is a surprisingly high level of ignorance as to the critical importance of making sure that customer satisfaction is achieved. The company has customers but so do people at each stage in the operations. The next downstream operation is the customer of the one before it. The same questions apply.

- Who are *our* customers?
- What are their needs?
- Do we satisfy their needs?
- How do we know?
- Have we ever asked?
- How much of the difference is determined by our suppliers?
- Do they know?

In many cases the least well trained and lowest paid workers are those that interface with the customer. This applies to high street stores and hotels but is common throughout the travel industry and the retail sector.

Pause for thought

- Ask yourself the question: Is that a problem in our company?
- If I were a customer how would I be treated?
- Phone the sales department pretending to be a customer and see what treatment you get. How were you treated by the telephone operator?
- Did you have to wait a long time either before the phone was answered or from there to speaking to someone who could help?

An organisation that conducted research of this question found to their horror that 7% of all callers hung up before speaking to anyone because they were not prepared either to go through a tedious routine of pressing this number followed by another and another before reaching a human being and then being entertained to some music that they may not have liked anyway. How many of those might have been potential new customers?

The following statistics indicate the scale of the problem. These have been collected from several sources on the internet and are a synthesis of data provided by some of the large global companies.

Winning and keeping customers

- It costs six times as much to gain a new customer than to keep an old one and 12 times more to regain one that has been lost!
- On average, one dissatisfied customer will tell 11 others who on average will tell 5 others.
- Often, an organisation's lowest paid people are the ones who meet the public.
- 96% of customers don't complain when they have a problem, they just don't come back.
- Half of those who say they are 'fairly satisfied' won't come back.
- The average company will turn over 10–30% of its existing customers because of poor service. Most could have been retained.
- Organisations giving quality service grow twice as fast and pick up market share three times quicker than their competitors.
- Companies giving quality service charge up to 9% more for their products or services.

For any strategy to be effective it is vital to obtain and analyse customer related statistics. In the context of this text, the word customer includes everyone who has an influence on current and future sales and pricing. This involves the end user, the retailers, stockists, relevant specialists, the media and the means of distribution. Clearly, sustaining customer satisfaction is vital to success, but this is a dynamic and volatile situation that requires constant monitoring and adjustments to the strategy.

Experience indicates that it is very dangerous to take advantage of the customer in a near monopolistic situation, thinking that the customer has no choice. Maybe they do not now but if things change? This was the case with state-owned companies and utilities. As soon as the monopolistic situation changed, people switched to their competitors at the first opportunity, even when the service might have been inferior, just to show their resentment at the treatment they had been forced to accept. This is a serious problem today for those countries that are attempting to liberalise state-owned monopolies.

Customer retention

It was claimed above that approximately half of the customers who say that they are 'fairly satisfied' do not come back. If the statistic is that it costs six times as much to get a new customer as it does to keep an old one, it makes sense to focus attention on the needs of existing customers especially if it costs nearly 12 times as much to get them back once lost. This is a dubious statistic. In practice, depending upon the reasons, it might be impossible to get them back. Also, there is the problem that a dissatisfied customer will tell 11 others who in turn will tell another 5.

Marketing and selling are costly activities, but referrals are not only free but potential customers are more likely to take notice of a referral than expensive advertising, which they do not trust anyway. It is a key question for the performance indicators. How much of our business comes from existing customers and how many are first-time buyers? Being able to sell at premium prices is another advantage of having a good reputation.

Customers want predictability. If they liked what they got before they do not mind paying a bit extra if they can guarantee that it will provide them with satisfaction. Of course, there are people who look for bargains and are more prepared to take a risk than most other people. At least one language teaching audio tape company sells the same product under more than one label. One is the name that is well known, the other is

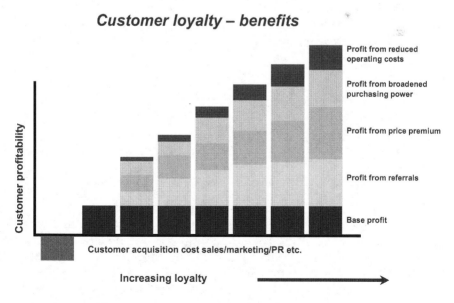

Customer loyalty – benefits

Profit from reduced operating costs

Profit from broadened purchasing power

Profit from price premium

Profit from referrals

Base profit

Customer profitability

Customer acquisition cost sales/marketing/PR etc.

Increasing loyalty

Figure 7.6   Customer loyalty

not well known and is sold through weekly news ads at bargain prices. They get the best of both worlds. Of course, the higher sales volumes, the more keenly can the price paid for component parts be negotiated and greater power and influence over suppliers.

Higher volumes also make economies of scale possible.

Summary to conclude the session

The customer is the key to growth. The leading organisations will be those who give the most attention to the achievement of customer satisfaction and customer retention. Statistics show that when 75% of customers say that they are fairly satisfied, 25% or more will try somewhere else. Getting an old customer back may sometimes be close to impossible especially when they have gone to what they regard as being a more customer-focused competitor.

Workplace learning

With your group, use brainstorming (explained in Chapter 4) or select a 'cause of customer dissatisfaction' topic from the process analysis chart:

- Brainstorm what you believe to be the causes.
- Write these on cards or Post-it stickers, one per sticker.
- What are your conclusions?

SECTION 5: QUALITY FUNCTION DEPLOYMENT (QFD)

In this section there will be several subsections, each progressively describing a tool used in QFD and following the path of our selected design requirement, the design of a radio-controlled racing car. You may decide to run ca project of your own alongside this one. It is a great way of learning.

Tools covered in this section:

- Affinity diagram
- Interrelationship diagraph
- Tree diagram
- The Kano model
- Prioritisation matrix
- Matrix diagram
- Process decision program chart
- Activity network diagram.

The first of the traditional QFD tools is the affinity diagram method. It is a good way to organise all the thoughts obtained in the previous session, which will be 'fussy' in the sense that they were identified at random, and systematically grouping them with related or like ideas hence the name 'affinity'. Here it will be described in its classic or normal form. Customer needs are many and varied and this is a complex issue. It can, and usually does, produce sometimes huge quantities of 'language data' that may appear seemingly impossible to deal with. Numerical data is normally generally much easier. Fortunately, the affinity diagram is an excellent tool to help with this.

The affinity diagram has many slight variants with different names because, since its origination in Japan by experts from JUSE (Japanese Union of Scientists and Engineers), it has been adapted and renamed by several people. For example, Professor Shiba from MIT (Massachusetts Institute of Technology) in the USA calls it the KJ (KJ being the initials of a Japanese consultant who created one version of this concept) method and refers to the inputs as 'fuzzy data' from his work on concept engineering and the technique is widely known by that name for that reason.

The affinity diagram method is used to clarify the nature of a problem, coordinate ideas, or obtain new concepts through the integration – based on affinity, of language data taken from a chaotic event or uncertain conditions.

Figure 7.7 shows how the affinity diagram method is a very effective way of bringing order out of chaos and to be able to take vague ideas and organise them in a coherent manner. In this example, the affinity diagram method has been used to identify the potential specification requirements for the design of a model racing car. However, note the use of the word 'potential'. It must be remembered that the affinity diagram has only ordered our thinking.

To create the affinity diagram, it is necessary to provide each member of the team with approximately ten Post-it type stickers of the size approximately 8 × 12 cm. Ask each participant to independently write on each of the stickers (one idea per sticker) the ideas that immediately come to mind that the customer will think important and that should be considered in the design, just as in our example in Figure 7.7. Expect between five and ten ideas per person but it is not a competition. When any individual is finished, ask them to go to a convenient wall space and stick

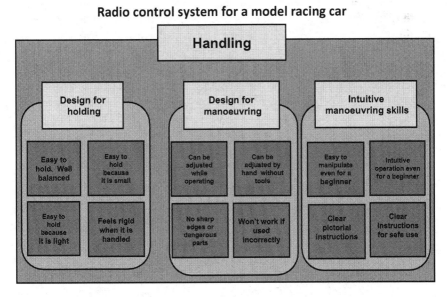

Figure 7.7   Affinity diagram – 1

Affinity diagrams – example
Radio control system for a model racing car

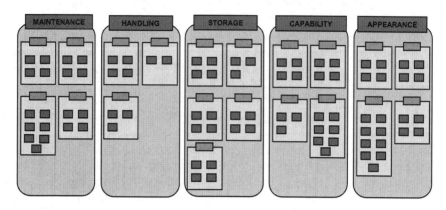

Figure 7.8   Affinity diagram – 2

their ideas there. As the ideas are presented, attempt to cluster them in related groups hence the term 'affinity'.

Depending upon the topic, there could be any number of groups but typically it could be as few as maybe three as in the example, or as many as eight or nine. If it looks like going beyond this then look at the ideas in more detail and attempt to combine some groups. Once this is done, then look at each group in detail and attempt to write a sentence that adequately describes the content of that group. As can be seen in Figure 7.7, one group there was labelled design for manoeuvring. The completed result should look something like Figure 7.7 but there is also an example in Figure 7.8.

In summary, the affinity diagram is a very effective method for first collecting the random thoughts of a group of people and then organising them in a coherent way in order to conduct further analysis. Note: sometimes it can take quite a long time to assemble all these ideas and it is rare, if it is done well, for it to be completed in a single session. If this is likely, then it is a good idea to pre-empt the problem and to put a large sheet of paper on the wall and make sure when all the ideas have been posted then at the end of the session carefully roll it up and store it until the next session. The tree diagram technique that follows enables us to structure and refine these ideas for further analysis.

Tree diagram

This technique normally follows the affinity diagram and works in the same way as the filing system on a computer. It works by progressively breaking a broad general aim into successively more and more detail until it has become several specific activities that cannot be broken down further. Figure 7.9 shows the method. Note that for the illustration we used the data from the affinity diagram in Figure 7.7.

If, as suggested, this technique follows the affinity diagram, the structured language data can be relocated using the same hierarchy. When the transposition has

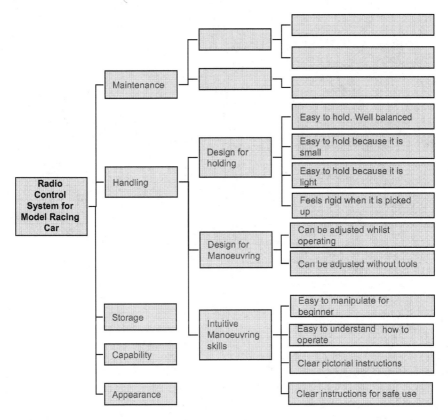

Figure 7.9   Tree diagram – 1

been completed, it will usually expose several deficiencies or gaps in the logic. These can be dealt with using this technique. On a large piece of paper (experience indicates that in most cases it is better to use paper to rough out the diagram before later transferring it to a computer programme) write the primary objective in a box on the left-hand side as shown in Figure 7.9 (for example, design and build a radio-controlled car). In the next column to the right, list each of the means that are available to achieve that objective and link them as shown. These 'means' then become objectives so moving again to the right, list the means of achieving each of these in the same way as before. Continue this process until it is not possible to break them down any further and this last column should now be very doable specific tasks or actions. The first step in using the tree diagram is now completed. Study the diagram carefully to see if there is anything that might have been overlooked. Do not rush this stage as it is important to include as much detail as possible.

The Kano model

So far, the affinity diagram and the tree diagrams have enabled the identification of needs and how they are structured. Prioritising them is important to ensure that time

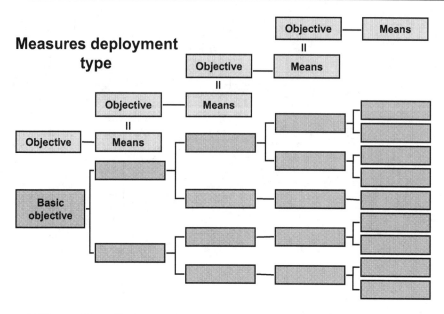

Figure 7.10   Tree diagram – 2

and effort are devoted to those that make the most impact on the objectives. The methods available are questionnaires and interviews, etc. The method that we would recommend here as a means of clarifying the **importance** of some of the key features is the Kano model. It suggests that customer requirements can be grouped into five main categories:

1. Attractive but not essential.
2. Must be.
3. Neutral – indifferent to it.
4. Can live with it.
5. Dislike.

In Figure 7.11 and using, in this instance, a popular family car as an illustration, fuel consumption would be an example of a 'one dimensional' feature. The lower the consumption the more it is liked. The braking system would be an example of a 'must be'. There is no positive 'customer satisfaction' from having efficient brakes but there would be considerable 'customer dissatisfaction' if it did not. A retractable radio aerial might be an example of an 'attractive' but non-essential feature. It is nice to have but is unlikely to be a factor in choosing the car unless everything else was identical to a competing model. Finally, there may be some possible features of which the customer is completely indifferent. This might be some extra button on the stereo system.

Using the Kano questionnaire, it is possible to group the various key features identified using the affinity method into their relevant categories, which will enable them to be assigned their respective importance.

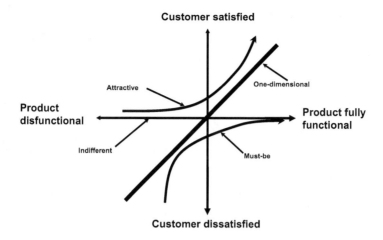

Figure 7.11   The Kano model

The Kano questionnaire

The questionnaire requires that each of the features to be investigated is considered in both its functional and dysfunctional form. Each interviewee must consider the items from both conflicting points of view. By doing this for each key feature and from a broad spectrum of potential customers of the product or service, the data obtained can be analysed.

By analysing each set of answers, the table in Figure 7.12 can be created, which will provide the QFD team with all the information they require to give the relevant weighting to each of the parameters identified.

Kano analysis

Interestingly, one DHI diploma student being trained by DHIQC used this technique to find out from a line of production workers what were the key issues that impacted on production performance. This unusual application produced some very revealing results.

Customer Requirement — **Tabulation of responses**

C.R.	A	M	O	R	Q	I	Total	Grade
1	1	1	21				23	O
2		22			1		23	M
3	13		5			5	23	A
4	6	1	4	1		11	23	I
5	1	9	6	1		6	23	M
6	7		2	3	1	10	23	I

Figure 7.12   Kano model data

Completion of the tree diagram

When the specifics have been analysed by whatever means, the tree diagram can be completed by adding three extra columns: 'feasibility', 'effects' and 'overall'.

Workplace Learning

If you made a start it would be a good idea at this point to complete the tree diagram including the prioritisation. If you wish, you may use the Kano model if only to realise how effective it is. For speed though, you may on this occasion use 'experience' to prioritise the features.

The house of quality matrix

This matrix is the central component of QFD and it is where everything comes together to create the final plan of the project. Refer back to the illustrations of the house of quality matrix as shown in Chapter 5 in the section on Hoshin Kanri. It shows a simple but typical construction of the house of quality. Note that there is a central matrix with additional 'rooms' to the top, right and bottom of the diagram. More of these can be added as required and there is considerable room for creativity in the design of a house of quality.

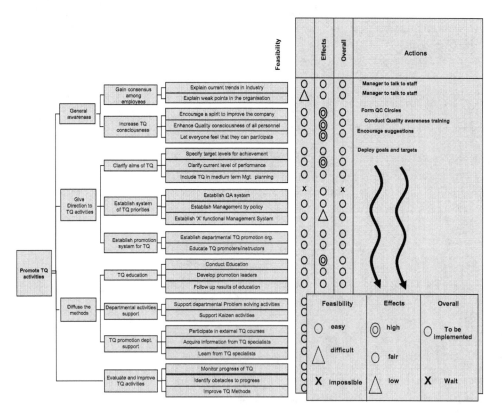

Figure 7.13   Tree diagram – feasibility

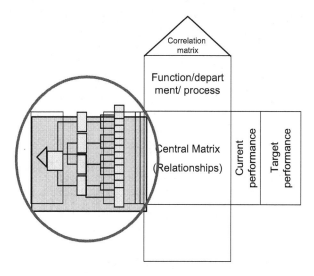

Figure 7.14   House of quality analysis

The application to new design was its original use and collectively the related tools from the affinity diagram through to completion of the new product development process the root. Note also that the tree diagram can located on the left-hand side of the central matrix. When the house of quality (HOQ) is used this way, it can be a continual monitor of performance against KPIs. If the chart is made large enough to cover a complete wall, it can be regarded as being somewhat like the British Prime Minister Winston Churchill's World War II War Room.

As changes occur in the market place, priorities will change amongst the performance indicators. When these are highlighted on the HOQ chart, attention can quickly be drawn to the new priorities and resources deployed. This will create an organisation that is capable of very rapid change. One of the main attractions of the house of quality is the fact that a very wide range of interacting variables can be seen on one diagram. In some organisations, all the key functions with their respective responsibilities. In the example in Figure 7.14, the matrix was used to log the ongoing performance of the KPIs and to initiate improvement projects where performance was below that required.

THE ARROW DIAGRAM

QFD projects that result in the need to implement a large programme with multiple complex events that all interrelate with each other may benefit from this technique. There are several variants, which include PERT (programme evaluation and review technique), CPA (critical path analysis), etc. but they are essentially all very similar, and we will use the more generic title 'arrow diagram'. Gantt charts are another alternative.

There are many different brands of software that cover this topic. Perhaps the most popular is Microsoft Project, which is easy to use, and which produces excellent results. The construction of an arrow diagram is simple if care is taken. Figure 7.15 shows the basic principles and terminology used in their construction.

Figure 7.16 (Arrow diagram – 2) shows how the chart is constructed for a simple operation involving two parallel streams of 'jobs'. The connection between nodes 3 and 4 shows how two activities on the separate streams are depicted when they must be performed at the same time. Just prior to each job, the numbers above the line indicate the node connections (e.g. 2–3) and the number below the line indicates the time that the job is expected to take which is usually but not always given in days.

Figure 7.17 shows that jobs on the lower path must be completed in the time allowed whereas those on the top path have some 'float' whilst the others are

Arrow diagrams

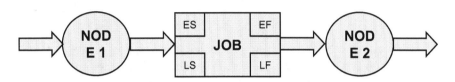

JOB – This is the activity that requires a length of time for completion.

NODE – This is the beginning and the end of a task and each is the connecting point to another job.

ES – Earliest start time possible.
EE – Earliest finish time possible.
LS – Latest start time possible.
LF – Latest finish time possible.

Figure 7.15   **Arrow diagram – 1**

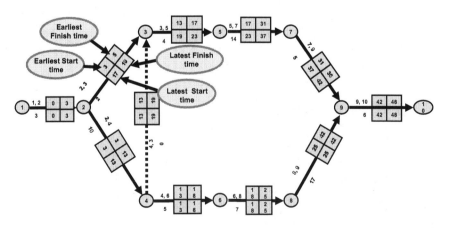

Figure 7.16   **Arrow diagram – 2**

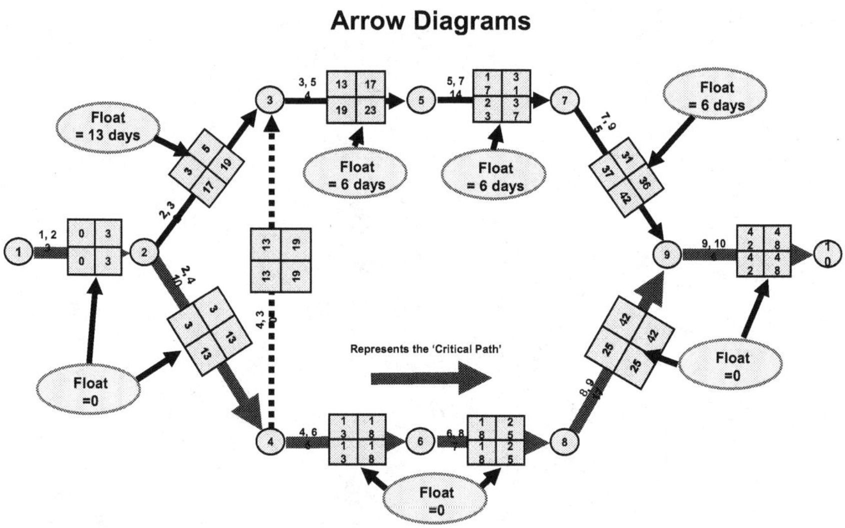

Arrow Diagrams

Figure 7.17   **Arrow diagram – 3**

being completed. This lower path therefore determines the overall time that it will take to complete the complete project. Whilst some delays can be tolerated in the jobs on the top line, any delay to any of the jobs on the lower line will increase the time for the whole project by that amount. This path, therefore, is called the 'critical path' and is emphasised by using a thicker line or, if possible, a red line. Conversely, if the time for the overall project is too long, it can only be reduced by reducing the time for jobs on that line. However, as this time is reduced, the float for each of the jobs on that line is also reduced. When this reaches zero then any further reduction in the overall time must also include a reduction in the time for jobs on the top line as well.

In a simple example such as in Figure 7.17 it would be easy to detect and deal with this, however, for a complex project it would be necessary to use a dedicated computer programme to make the calculations.

Even with the power of the techniques of quality function deployment there is no guarantee that every feature of the plan will necessarily work without a hitch of some kind. In fact, the more innovative the plan is the more vulnerable it will be to a failure of some sort. Risk management goes into this in some detail and the technique failure modes and effects analysis (FMEA) can be used to analyse the features where the risks might be high or their realisation being severe to catastrophic. However, there is another technique that is popular in QFD, which is known as PDCP (process decision programme chart).

Process decision programme chart (PDPC)

This technique is somewhat like fault tree analysis, which is itself a variant of the FMEA approach. The following diagram is based on a similar example in Japanese

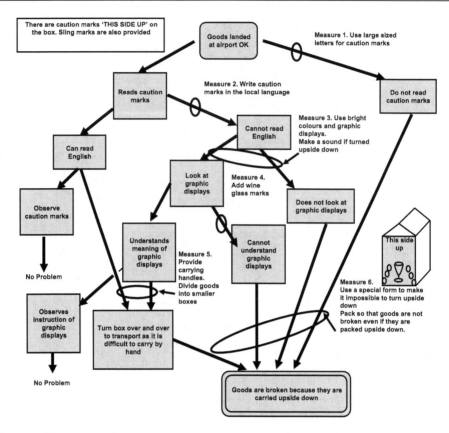

Figure 7.18   PDPC

training materials and shows a simple PDPC chart for the delivery of a package by air to Heathrow Airport. The chart has been designed to try to eliminate the risk of a package being delivered 'upside down' due to the fragility of the contents.

The process begins with the obvious decision to mark the box 'This side up'. This might seem a good idea but is it fool proof? It would appear not, for two reasons. 1) The handler may not read the message or 2) He reads it, but this leads to two more possibilities – first, he may not be able to read English or, second, he can read it and observes the caution, in which case no problem, or, alternatively, he still turns the box upside down because it is easier to carry.

By following the alternative routes and the precautionary measures that have been taken to deal with the possible outcomes, in the end the package was designed in such a way that it was impossible for the package to be delivered upside down. This is the essence of PDPC and *poka-yoke*.

Interestingly, this technique was introduced by DHI to members of the accounts department of a large steel works. They were in the middle of designing a new pay scheme for the entire organisation that would eliminate direct payment by results. Normally, the introduction of a companywide scheme of such proportions would be fraught with problems and disputes. Amazingly, by using this technique to pre-empt and deal with the most likely problems, the scheme was introduced with virtually no issues from anyone.

Failure modes and effects analysis (FMEA)

Definition

FMEA is used to evaluate and predict the effects of a potential failure mode in terms of customer performance or perception.

Potential effects of failure

- Doesn't work
- Intermittent operation
- Doesn't fit
- Appearance of final system affected
- Unstable performance
- Discontinuity
- Short circuit
- Field failure
- Consumer recall or returns
- Reliability in question
- Compromise customer's reputation
- Difficulty assembly
- Doesn't meet capabilities of the competition
- It is used to recognise and evaluate potential failures when designing and manufacturing a product
- It identifies actions that could eliminate or reduce the potential for failure
- It documents the process
- It is a systematic approach for product/process reviews
- Helps define customers' needs and expectations prior to final design
- It is meant to be a 'Before-the-event' action
- We always have to find time to 'Do-it-again' but we never have time to 'Do-it-right' the first time
- We will never improve our product information
- Evaluates design requirements and alternatives
- Potential failure modes are considered before actual design takes place
- Helps define test program and customer acceptance
- Ranks failure modes in terms of customer requirements (this allows proper priorities)
- Allows mechanism to analyse field concerns design alternatives
- and in the case of components or parts:
  - All changed parts
  - New application of existing parts
- Avoids duplicating another design we cannot manufacture.

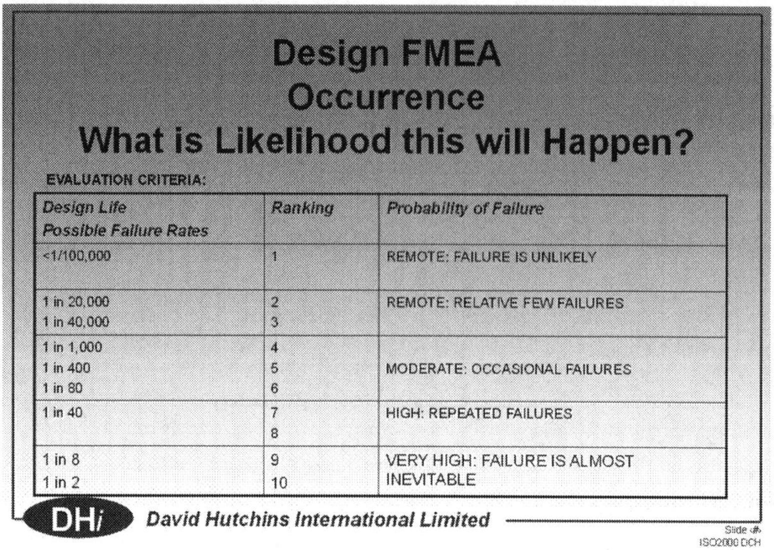

Design Life Possible Failure Rates	*Ranking*	*Probability of Failure*
<1/100,000	1	REMOTE: FAILURE IS UNLIKELY
1 in 20,000 1 in 40,000	2 3	REMOTE: RELATIVE FEW FAILURES
1 in 1,000 1 in 400 1 in 80	4 5 6	MODERATE: OCCASIONAL FAILURES
1 in 40	7 8	HIGH: REPEATED FAILURES
1 in 8 1 in 2	9 10	VERY HIGH: FAILURE IS ALMOST INEVITABLE

Figure 7.19 FMEA – 1

Design verification

Design verification is a method to predict and determine failure will occur prior to product release to the customer.

Detection

Detection is assessment of the ability to identify a potential design weakness.

Methodology

- Product testing
- Design reviews
- Prototype tools and tests
- Finite element analysis
- Computer simulation and modelling
- Any other mathematical studies.

Risk priority number

This is a calculated number showing the relationship between three aspects of risk.

1. S – Severity of failure.
2. O – Occurrence or frequency of failure.
3. D – Ability to detect failure.

Design FMEA
Detection
Probability of Detecting a Weakness

EVALUATION CRITERIA:

Ranking	*Likelihood of Detecting a Potential Failure*	
1 2	VERY HIGH	Testing will almost certainly show potential design weaknesses. Function simulation is excellent
3 4	HIGH	Program has a good chance for detection, Function simulation is weak, but two tools for verification are used
5 6	MODERATE	Program may detect failure Functional simulation not feasible
7 8	LOW	Program likely not the detect failure No simulation or functional testing
9 10	VERY LOW	Program probably will not detect potential design weakness. Customer application unknown or no test performance in this area.

DHi *David Hutchins International Limited*

Slide 147
ISO2000 DCH

Figure 7.20 FMEA – 2

Process failure mode and effects analysis

- Evaluates the potential product-related process failure modes
- Identifies the potential customer effects of failures
- Identifies significant process variables to focus upon controls and reduction of risk
- Ranks failure modes in terms of their effect on the customer
- Relates the product to the process and the impact of the process to the integrity of the product
- Normally the customer is the 'end' customer but also includes the subsequent downstream manufacturing operation
- Summary of tools that are popular in QFD
- We do not cover process analysis in this chapter as it is more closely associated with business improvement techniques explained in Chapter 4.

Just six of the techniques of quality function deployment have been included in this chapter but there are many more. These include Gantt charts, design of experiments, cause relationship diagraphs, dynamic programming and linear programming and many variants of those that have been included. It is recommended that the student becomes familiar with all of these powerful techniques in order to get the best benefits from quality planning.

SECTION 6: BASIC STATISTICAL RELIABILITY

What follows are some basic skills that would normally be found in the functional design department. The word 'reliability' can be the 'life aspect of quality' and this is

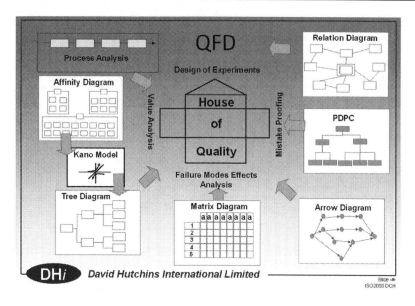

Figure 7.21   House of quality tools

one of the prime tasks of that department. It is not just useful in engineering manufacture, there are applications everywhere depending upon need.

For example, in the manufacture of furniture it is not unusual for the manufacturers to set up a rig to drop heavy weights repeatedly on to sofas, chairs and beds, etc. to determine their fatigue life. There are rub tests on cloth for the same reason. Occasionally, you can find strange sections of carpet on heavily used walkways at airports etc. to check on wear.

In this chapter we only include an appreciation of the basic concepts. The objective is that the relevant executives and managers should be able to ascertain whether the relevant activities are in operation. Unfortunately, experience indicates that, for commercial reasons, usually related to attempts to rush a new product or service into the market place, this vitally important activity is often either bypassed altogether or only treated with lip service. It is often for this reason that tragedies like the bridge collapse in the USA in early March 2018 occurred only one week after it was installed, the failure of the Challenger space shuttle and events too recent to be tastefully mentioned in this book but which are in everyone's memory.

Here is an excerpt of a report on one such key event:

> Blinded by the success of the early Shuttle flights, the Agency's management had developed a careless attitude towards warnings coming from the engineering community.
>
> NASA had committed the Shuttle to an impossible schedule even before it entered service in order to ensure funding.
>
> Over time, NASA management had grown increasingly impatient with the technical delays that operating such a complex machine required.

All that ended on the bitter cold morning of 28 January 1986, when seven astronauts lost their lives in front of family, friends, and millions of TV viewers. A vehicle that was celebrated for its technical prowess broke up 73 seconds into the flight, burning nearly 2 million litres of fuel in just a few seconds, creating a sinister cloud of gas that still plagues the memory of anyone who saw it.

At this stage of the design process we are concerned with risk. Risk is always present in every situation and it will not diminish to zero ever, but we can do everything possible to attempt to minimise it. At this stage of design, we want to a) be able to identify what the risks might be, and b) to reduce them as far as possible. This requires product reliability testing and data analysis. In many cases this may be simple and not take a long time but in others, especially with complex systems where there may be multiple failure modes and interactions, it can be highly complex and well beyond the scope of this book. Only those with PhDs that are dedicated to this field need apply!

Whilst FMEA (discussed earlier) is a predictive reliability technique, we also need to verify the actual capability of designs through prototype testing. Generally, this is carried out on production standard prototypes and in environmental test laboratories using a range of statistical tools. One of the most popular of these is the technique called Weibull analysis but there are many others. We have included some examples to illustrate how Weibull analysis might be used and its value in design validation. One of the popular features of Weibull analysis is that it requires very little data. This is useful when the test is destructive or takes a long time to perform.

When equipment is tested to verify the specified quality or reliability characteristics testing can only realistically be carried out using statistical sampling techniques. This is because it is usually impossible to test the entire population. Samples may be obtained from the following:

(a) Random samples taken from the population.
(b) A sample of time taken from the life of the product.
(c) The time taken to obtain a predetermined number of failures.
(d) The number of trials to obtain a pre-selected number of failures.
(e) A combination of a, and b or c.

The population might consist of any of the following:

(a) The life of a single product.
(b) Development models.
(c) Pre-production runs.
(d) Production runs.

As with other forms of sampling, care should be taken to ensure that the items selected for test are as representative as possible of the population.

Assumptions

Unlike many other statistical applications in industry, we are unable to make assumptions as to the underlying-distribution relating to our test data. The 'normal

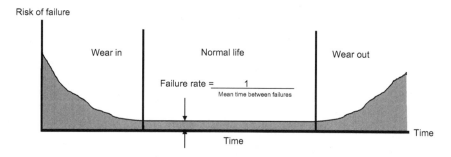

Figure 7.22   Bath tub curve

distribution' that many people are familiar with and that is extremely common, is only one of several possible distributions that will be found when reliability testing. In fact, it is only common when dealing with 'wear out' failures. This is another advantage of Weibull analysis, because to use it the underlying distribution is not necessarily known but the results of the analysis will show which they are.

Equipment failures may relate to several different distributions, and it is often necessary to determine these before calculations as to failure rate, mean time between failures, or confidence limits on reliability related to time can be made. The following curve, frequently described as the 'bath tub' curve, describes the life of a typical electro mechanical product in terms of risk of failure (or hazard) plotted against time.

The curve shows three distinct zones.

1. **Wear-in period** – in this period the risk of failure is decreasing with time. This is also known as the 'burn in' period.
2. **Normal life** – during this period only random failures may occur. The length of this period can often be extended by good maintenance. In some cases, theoretically, once past the 'wear-in' failure stage, there are products that apart from the occasional random failure will live on indefinitely. This is the case with many solid state electronic products. They will fail but often it is because of the knock-on effect of the failure of some other part.
3. **Wear-out period** – the end of the useful life of the equipment which may then be either-overhauled or replaced. It is the way failure occurs in this period that will help determine the appropriate maintenance/replacement strategy.

Reliability prediction

Reliability data usually refers to the normal working life section of the 'bath tub curve'. During this period, typically failures only occur at random, i.e. a Poisson failure distribution, and are usually due to unpredictable causes, such as stress exceeding strength or maybe a previously undiscovered weakness in an individual component.

Probability of item failing in next small interval of time is constant = (Lamda λ) per unit time.

where λ = hazards (in exponential case only) failure rate.

Typical units are:

- failures per hour
- failures per 1000 hours

Poisson distribution for events occurring in a random manner.

$$P(r) = e^{-m}\, \frac{m^r}{r!}$$

for constant hazard, in time t, we expect λt failures i.e. m = λ t

where P = Probability
r = no of failures
m = expected no of failures in time t.

Probability of zero failures in interval 0 λt

$$= e^{-\lambda t}\frac{(\lambda t)^0}{0!}$$

but the probability of zero failures = Reliability = R

$$R = e^{-\lambda t}$$

Series systems

System only survives if a, b, c and d survive

Example:

Engine, gearbox, transmission and differential of a motor car, i.e. not necessarily a system in the electrical sense.

$$R \text{ system} = Ra \times Rb \times Rc \times Rd$$
$$\text{and } R(t) = Ra(t) \times Rb(t) \times Rc(t) \times Rd(t)$$

Example:

Given the failure rates of the following components, find the meantime between failures and the overall system reliability.

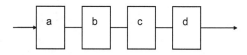

Figure 7.23   Series systems

Component type	Number in use	Failure rate %/1000hrs	Overall failure rate %/1000hrs
Valves	150	0.029	4.35
Resistors	90	0.1	9.00
Capacitors	110	0.1	11.00
Coils	20	0.1	2.00
Sockets	10	0.3	3.00
Transistors	80	0.1	8.00
Contactors	4	0.5	2.00
Total			39.35

Figure 7.24   **Failure data**

$$\text{Failure rate} = 39.35\%/1000 \text{ hrs}$$
$$\text{Meantime between failures(MTBF)} = 1$$
$$0.0003935$$
$$= 2541.3 \text{ hrs}$$

Parallel systems (redundancy)

The system fails if A **and** B fail. It survives if A or B survive, i.e. active redundancy – no switching involved.

$$\text{Probability that A fails} = 1 - Ra$$
$$\text{Probability that B fails} = 1 - Rb$$
$$\text{System probability of failure} = 1 - Rs = (1 - Ra)(1 - Rb)$$
$$\text{Let } 1 - R = Q \text{ (unreliability)}, 1 - Rs = QA \times QB$$
$$= (1 - 0.81) \times (1 - 0.81) = 0.0361 = \text{System probability of failure}$$
$$\text{Therefore, probability of survival} = 1 - 0.0361 = 0.9639$$

Reliability testing

To maximise the likelihood of avoiding the types of catastrophic failure mentioned in the introduction to this section, it is necessary to carry out exhaustive testing at the

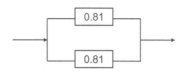

Figure 7.25   **Redundancy**

design stage. Unfortunately, with many products the time to reach the 'wear-out' phase can be years with no prediction at the outset as to how long that time might be.

There are several possible strategies here but the most popular are:

Accelerated Life testing

In this case, the severity of the loads, pressures and stresses that might occur in normal life are increased with multiple repeated stress reversals continuing to failure. This is intended to determine both the failure mode and its likely consequences as soon as possible. The down side of this approach is that the test method itself might create an artificial failure mode that might not occur in normal use. For example, the rapid reversals might induce a heat build-up that might not occur normally.

Duane model

This is popular in the design and build of both aircraft and automobiles, in this case whilst the usual prototype tests are carried out at the design stage. Because of the potential flaws in accelerated life testing, when it is thought that the design is robust, several prototypes are put into simulated use ahead of the normal production models. These special prototypes are subjected to all the typical stresses of the production units but always ensuring that they have many more accumulated service miles than any production unit. These special prototypes are vigorously inspected periodically. The intention is that if there is an inherent failure type then it should hopefully make itself obvious before it occurs in normal production models.

In both types of testing a popular technique is Weibull Analysis. Weibull analysis can be done mathematically but the maths is difficult as it was originally derived empirically and, fortunately, there has evolved a simple graphical approach that can be practised by using some of the specialist's texts on this advertised on the Internet.

Environmental testing

This involves testing the product or service features in the real world. The idea is to simulate as far as possible real-life situations to see how the product behaves or if the service does what is intended. It is better to learn this before product launch than afterwards! In the case of automobiles, this might involve driving the vehicle in extreme conditions, for example, in the Arctic, the tropics, across deserts, through mud and sea water, testing them with tall or short people, young or old. Similar ideas go for aircraft and this sort of testing is carried out by test pilots who push the product to its boundaries. Occasionally, there might be a tragedy but, fortunately, the checking is usually meticulous and potential failure modes are discovered before any harm is done. Then there are environmental test laboratories. These, again, are designed to put a product through its paces, with shock treatment, sand blasting, thermal shock, vibration etc. Techniques such as Weibull analysis and FMEA, etc. can be used in both life testing and environmental test situations.

Pause for thought

If you have studiously worked through this longer chapter, you should be well placed to check the methods used in your design office, operations and customer-related product or service information to look for possible opportunities for improvement.

8 *Supply chain management*

What you will learn in this chapter:

- Advantages and limitations to customer and supplier of multi-sourcing short-term supplier relationships compared with near single sourcing long-term supplier relationships
- The risks concerned with disruption of the supply chain
- Overview of the respective roles of the quality and purchasing functions
- Trends in supply chain management due to globalisation and computerisation
- Supplier monitoring
- Comparison of vendor rating schemes, the use of 'preferred supplier' status and other supplier motivation schemes such as 'supplier days' joint advertising, etc.
- Incoming goods control, supplier traceability
- Plan a pre-audit of a potential supplier based on the fictitious 'Cerif' case study.

SECTION 1: INTRODUCTION TO SUPPLY CHAIN MANAGEMENT

It is not at all uncommon for suppliers to complain that their customers have one set of rules for them and another for themselves. This might sometimes only be a gripe, but it is logical that if there is such a thing as 'best practice' then why expect your supplier to follow it and not do so yourselves and vice versa? The best way to look at it is to consider your supplier's processes as being an extension of your own. Ideally, there should be no obvious join.

With some 60% or more, sometimes almost 100%, of the component parts of manufactured products being outsourced and with large companies having as many as 40,000 suppliers, sometimes even more, supply chain management is one of the most important aspects of modern quality management. Before we start. Let us first be aware of a very important fact. There still exists in many countries two distinctly different approaches to supply chain management.

SECTION 2: MULTIPLE SOURCING

The first approach, which originated in the West, is referred to as 'multiple sourcing'. In this case, a customer will try to find several possible sources for a given product or service and try to force them to comply with contractual requirements with the threat of the loss of future business. This has proved to be both a short-sighted and seriously flawed strategy. Sadly though, it is still commonly being used.

Pause for thought

Check out your own organisations strategy in this respect. Is it flawless?

In the West, until the 1980s, this was almost the only system that organisations used, and it was thought to be good business, at least from the customer's point of view. It was very common especially where high volume turnover was concerned. In recent years this philosophy has been severely challenged by post-World War II Japan. In their system, long-term collaborative relationships are established, which in the case of Toyota are referred to as the 'Toyota Family'.

Let us look at a case example of the 'Toyota Family' approach, which will hopefully illustrate why multiple sourcing has been largely discredited.

SECTION 3: WHY SUPPLY CHAIN MANAGEMENT (SCM) IS CHANGING

This example is selected because it brings out some of the key points as to why SCM is changing. Hopefully, it is sufficiently graphic to make sense to everyone whether involved in manufacturing or not. The principles highlighted are applicable to all industries whether service or manufacturing. Unfortunately, in a service environment, examples that might be relevant are not so visually dramatic.

The example is from one of five major Toyota passenger car assembly plants in the Nagoya City locality in Japan between Tokyo and Osaka in the south. This visit took place in 1980 before many Western observers had seen such concepts as just-in-time assembly, single minute exchange of dies (SMED), *jidoka* or any of the other features of what has since been labelled in the West as 'lean production'.

At first sight the plant looked much the same as any other car assembly operation. However, it soon became apparent that there were some obvious exceptions. First, there were live pot plants along the assembly track. Then we noticed two-metre-wide carpet along the track. The workers wore soft shoes and the lighting was brighter than was typical in the West at that time. The workers all looked smart wearing clean overalls, as did the supervisors and managers.

Figure 8.1   Visitor data

It was clear from the number of coaches in the car park when we arrived that we were not the only visitors, there must have been hundreds of others. We were then very surprised to discover that the majority of these were school children aged about 10 to 12 years.

In order that they and we did not interfere with the production process, we ascended a metal staircase at the beginning of the line and walked along a gangway running the length of the track about 10 metres above it. The gangway was full of the young children all with notebooks making observations of the production process. It was clear that this happened daily. Looking down at the assembly operation, the first significant difference between this and equivalent operations in the West was the almost total lack of inventory at the trackside. Behind the track there was a road where forklift trucks were constantly running back and forth with small quantities of parts for replacement of the small stocks at the trackside. Bear in mind that this was 1980, it was well established, and this was some 15 years before the book *The Machine that Changed the World* was published in the USA and which coined the use of the term 'lean production'!

Partly assembled vehicles slowly moved along the track where assemblers were stationed approximately every 10 metres or so. As the next vehicle approached an assembler, he had just completed his task on the preceding vehicle and, with no measurable pause, he would take the appropriate parts from a rack that had just been replenished by one of the forklift trucks and begin to assemble them into the next vehicle. This was the same all along the line, all the workers were working at the same pace, no rush, no waiting, everyone appearing to work in unison.

Figure 8.2   School visit to the Toyota plant

Suddenly, there was the sound of a siren. Looking down the track there was a yellow revolving light flashing over one of the workstations. The line stopped immediately, and three workers left their stations and ran to where the light was flashing. It became clear later that the worker in that area had run into a problem in his work. Therefore, he had pulled a cord above the workstation that stopped the line immediately and set the light flashing. We also learned that the three workers who ran to his aid were members of a quality circle of which this worker was also a member. In a few moments, they had solved the problem temporarily and returned to their stations. At that point the worker who had the problem pulled the cord again, the light went off and the line restarted.

Our attention was then drawn by our guide to an illuminated rectangular screen (as illustrated in Chapter 3) above the production line, split into quadrants, each including a three-digit number. The top left number indicates the target number of completed vehicles for that shift. The top right number indicates the number of vehicles that should have been completed by this time if the target is to be achieved. Bottom right is the number that have been completed. Bottom left is the number of minutes lost since the start of the shift for stoppages such as the one we observed.

We were told that at the end of the shift, the workers in the relevant quality circle would meet and attempt to find a solution to the problem so that it cannot reoccur. This means that every day, day after day, the line becomes progressively more and more efficient.

We were informed that this concept of stopping the entire line just for a single small problem is called *jidoka*. At first sight it might seem expensive to stop the entire track just for one simple problem and, of course, it is, but not anywhere near as expensive as the alternative of muddling along and not building a record of where the real problems and where the real costs are hidden. Had the line not been stopped, then there would have been an accumulation of so-called 'work in progress' upstream from the location of the problem. The term 'work in progress' is a misnomer because, in fact, it is not progressing, it has got caught up in a log jam. Downstream, the work would stop or slow down anyway being starved of product whilst the problem is being solved. That loss of time is unlikely to be recorded; the problem will not be solved so it remains inherent in the process and will reoccur repeatedly along with a myriad of other problems. All of these will result in a build-up of so-called 'work in progress' as well. Eventually, overtime will have to be paid to clear it but still the problems will remain.

Figure 8.3   Japanese QC circle

Further along the track we were surprised to note an open-backed truck backing up to the line and we now come to the core point of this case example. The truck was loaded with engines. One of the engines was being slung with ropes and lifted by a hoist out of the truck and lowered into the engine compartment of the vehicle being assembled. Four workers were involved with that and immediately it was located they began to fix it in position. Then we made another observation. The engines were not alike! We then noticed that the vehicles on the track were also different from each other. In fact, there were several different models mixed along the line.

The engines were loaded in the truck in the same sequence they were required according to the vehicles on the track! This meant that the engine supplier must have known in advance the scheduling on the track, the precise time that each vehicle would be at that assembly point and loaded the truck accordingly. We also noted that as the truck became empty, the next truck had arrived and was manoeuvring into position.

This raised several questions in our minds. Were there many trucks all waiting in the yard? Did the engine manufacturer stock large numbers of engines so that he could pick and mix on the day? What would happen if a truck failed to arrive? What if it had been loaded incorrectly?

We were then told that the trucks came directly from the engine manufacturer and that, no, he did not have any stocks of engines. All the engines were manufactured exactly to production requirements, tested and shipped with no interruption. We were also informed that they had not had one single stoppage due to faults in this process in the previous six years!

Further along the assembly track we saw the same procedure with tyres. All of this is the essence of what has since been labelled 'just in time'. This concept challenges many of the long-held perceived principles of production and the widespread Western belief in the apparent logic of 'economic batch size' and 'economies of scale'.

Later in the visit we also saw the operation where the body parts are spot welded automatically onto the chassis. This is a complex caged area with a large array of robotic arms that search and pick the relevant parts, holding then in place whilst the automatic spot welder stitches them in location. After the completion of just two bodies, suddenly the top of the cage opened, and many of the robotic arms instantly ascended and were replaced by others.

The door closed, and the next body was welded. The changeover happened in much less than one minute. In a Western automobile manufacturing plant at that time, such a changeover would normally have taken a complete shift. This concept is now known as SMED (single minute exchange of dies) but, again, this was 1980!

In Western thinking, rather than challenge the fact that 'economies of scale' are only economic where there are unavoidably long set-up times, these set-up times were never challenged but considered to be a necessary fact of life. The Japanese manufacturers challenged this thinking, realising that if changeover times were zero then you could mix and match production at any time depending upon need. It was another 15 years before Western auto manufacturers realised and responded to that. Instead, they took those set-up times to be a fact of life and, instead, attempted to spread that high cost over as many parts as possible. Not only did

Toyota challenge that but we saw exactly the same phenomenon at Nissan on the same trip. All leading Japanese manufacturers challenged this belief with the stated consequences.

This meant that they could also produce vehicles according to demand instantly. Theoretically, this meant that a potential customer could make a choice in the show-room; the vehicle might not even exist but as soon as the order was placed it could be manufactured almost immediately. Later in the process, near to the end of the line, we noticed that specific individual customer names and requirements were posted on A3 forms attached to the windscreens. Importantly, this type of achievement is not possible if multi-sourcing is used because the entire process requires total harmony right through the supply chain.

The following day we visited the engine plant, which was only a matter of a few streets away. On another occasion we went to a Toyota supplier gear box foundry, a plant where cooling pumps were manufactured and NGK sparkplugs. In every case, there was no inventory, and everything was specifically destined for a vehicle already in the process of manufacture.

Also, defects cannot be tolerated. A single defect either in the engine or any of its component parts, or indeed in any other of the 2000 or more components, can rack up huge disruptive costs by line stoppages waiting for replacements, etc.

Obviously, a production system like this could not have been created overnight. It took these Japanese companies from 1950 through to 1980 before it was perfected and still it continues to be developed. Even to this day Toyota claim that in their Japanese plants from a labour force of some 40,000 people they have over 2.6 million improvements per year of which 96% are implemented! Compare this with typical practice even to this day in the West and it can be seen why even after three decades of attempting to catch up, the gap is still so large.

Pause for thought

What lessons might your organisation learn from this example? OK, maybe you do not produce automobiles but consider the principle. How do you manage this aspect of your supply chain? Consider the principle of mutually beneficial customer/supplier relationships.

Where does your company stand on mutually beneficial customer/supplier relationships?

It makes sense that 'best practice' should apply to both supplier and customer alike! Is this true for you? Good supply chain management should be a *mutually* beneficial process as can readily be seen from the Toyota example above!

Even after two or possibly three decades of watching companies succeed with this policy, many Western organisations still operate the outmoded and totally discredited system of multiple sourcing and adversarial relationships with suppliers based on the maintenance of a fear culture. How about your organisation? These companies can only get away with that for as long as their competitors continue with similar policies.

SECTION 4: ASSESSING THE ROLE OF STRATEGY AND BUSINESS OBJECTIVES WHEN SELECTING AND MANAGING SUPPLIER RELATIONSHIPS

Strategy and business objectives begin with **people**. If people do not care, if they just do their jobs according to instructions, then it does not matter how much **system** you have, you will never achieve your goals.

First, you have to attempt to create an environment where everyone is working towards making their organisation the best in its field. So, before we start talking about strategy and business objectives, we will quickly review some of the **people** issues. Is this the case with your suppliers as well?

Pause for thought!

So, the question remains: is your organisation or your supplier organisations based on the Taylor model?

If so, what problems might you face? Are yours and their employees happy? Are their opinions sought after? Do they look forward to coming back to work after a weekend?

SUPPLIERS' MANAGEMENT SYSTEM MODELS

In terms of the number of implementations around the world (but not necessarily in terms of effectiveness), ISO 9001 is the most successful management model with registrations in over 100 counties in both manufacturing and service industries. However, it is not without its critics who are many and some who are quite vociferous! We recommend that you pay attention to some of the critics!

It has also been the basis of many industry specific standards that have extended and modified wording to suit the industry. In some instances, there have been additions to clarify the requirements.

For example, within the food industry it is a requirement to have a system of food safety. One such food safety system is hazard analysis and critical control point (HACCP) as explained in Chapter 6. Although within ISO 9001 it states that the organisation must meet its appropriate statutory regulations within ISO 22000 (the food standard) there is a whole section on HACCP and its requirements, which is just one of the food industry regulations.

The question is, just because our suppliers might be certified to ISO 9000 or, for that matter, any other standard, does this necessarily mean that they can confidently be trusted to meet our requirements such that we can depend upon them in the same way that Toyota can depend on its suppliers? Sadly, definitely no. Not at all. For one thing, these standards are related to management systems and not specific product performance.

We recommend, once again, that the reader should take a look at the UK Trading Standards website on the subject of product recall covered in Chapter 1 for a different reason than mentioned in earlier chapters. The top 50 most recent recalls are published there. If you scroll down them and read the accounts of the mishap, two things will stand out. First, a large number of the recalls will have their roots in product design and further investigation will indicate that a large number of them will have been certified to ISO 9000 as the vast majority of companies are. I have pointed this out repeatedly over the entire life of the standard and its predecessor BS5750 and I am always greeted with a shrug of the shoulders and the comment, well that is not what the standard is there for. To date, nobody has convincingly told me exactly what it is there for!

All I know is, that if my business depended upon the services of suppliers, I would want to know at least the following:

- Do they have the plant and equipment together with measurement systems that give me confidence that they have the capability to meet my requirements? In certain critical features I would want to test that capability first hand by carrying out gauge R&R, and statistical process capability studies on key plant and equipment.
- How are their employees treated? Is it an autocratic management style or participative? If they claim to have quality circles, I would want to talk to them and check how they are managed, the training received and whether they are authentic.
- I would want to know if the organisation was financially viable. I would not want them to get into financial trouble whilst they were serving our needs.
- I would want to know about potential conflicts of interest between ourselves as a customer and other interests of the supplier. We would not want either to be compromised or to have to play second fiddle to anyone.
- What is their reputation in the industry do they bring references?
- If things go wrong do, we have confidence that they will deal with it efficiently and appropriately.
- Do they have good knowledge of and practise the concepts included in this book?

Given all of that, then they would probably qualify for preferred supplier status, but I would keep a close eye on their performance for maybe the first year. All being well, after that, if we have a very large number of suppliers, I might give them the opportunity to manage a collection of them as I might others at the same level.

All of these would be offered long-term relationships and would be encouraged to share ideas, technology and methods with both us and each other for the mutual good. I do not think any of this can be found in any useful form in any standard that I have seen to date.

So, what can be done? The answer is that if you intend to have a seriously effective supply chain management process that runs seamlessly into your own, then you simply cannot rely on any form of third party certification. It is something that you have no choice but to do yourselves. This is known as second party auditing and you cannot use any pre-created standard to do it. The reason being that what you are looking for might not even be mentioned in this or any other standard. Your needs, your ideal processes, will be unique to you. What you create will suite your organisation and yours alone. You will need to have the courage of your convictions. Look to see what others are doing but if they are doing something and you are not then you should ask yourself why? Maybe it is not relevant to you. The converse is also true. Maybe you have found something important that will take you forward and you cannot find anyone else doing the same. Well, have the confidence of your convictions. This is what leaders are made of. You could end up being a trend setter and the Toyota or Nissan of the future! Who knows.

Another important point is to check every one of your supplier's processes for evidence of Dr Shewhart's PDCA cycle. Are all the loops closed? If not, then what might be the consequences?

SECTION 5: SUPPLIER PEOPLE POLICIES – DETECTING TAYLORISM OR THEORY X MANAGEMENT

In the worst-case scenario, you may discover in effect that each department of the supplier performs its activities and then in effect 'throws the output over the wall' to the next department. There would be very little communication in either direction between the two and virtually no understanding of each other's needs. When things inevitably and frequently went wrong, the practice for each would be to blame the other. A blame culture in easy to detect and difficult to hide. It might be obvious by what people say, but often body language is a giveaway.

Some managers would be more expert at the blame game than others, consequently they would appear to be the most competent and therefore the ones most likely to be promoted. Since this was how everyone was promoted, they then found themselves confronted with a better class of enemy! The nearer they got to the top of the organisation, the tougher and more sophisticated became the competition.

At the highest levels intimidation was the main weapon. The most successful managers would eat in a separate and more luxurious restaurant, they would have large offices, deep pile carpets, oak panelled walls, named car park locations, dress differently and never speak directly to the workforce. Such distinctions are regarded as 'perks' but, in fact, are really a form of intimidation.

This method of management could be regarded as the modern equivalent of the North American Indian's war paint! Companies that pride themselves in more participative styles of management scorn these distinctions and make great efforts to avoid them. I mention it because it is this type of supplier that is likely to present the worst problems with the blame game prevailing.

Whilst the worst extremes of this autocratic approach have thankfully largely become less obvious in recent years, it is, nevertheless, often still lurking beneath the surface and few managers are either trained or encouraged to use participative management techniques.

We are therefore still faced with the worst in macho-based autocratic management, and no wonder so many executives and managers have stress disorders and heart attacks in their late 30s and 40s! It happened to a friend of mine. He suffered a nervous breakdown but before that was being sick when he got up in the morning before coming to work just by the thought of it. Hopefully, your management style does not have that effect! Whilst autocratic management methods may appear to have receded in recent years and are being regarded as being 'politically incorrect', they are, nevertheless, proving to be self-sustaining and very resilient.

Those executives, managers and who do reach the top in such organisations, do so because they are good at this style of management. They know how to make the method work in their favour and are not interested in experimenting with any alternative that might make their position less secure. Their belief is also strengthened by their observations of the behaviour of those beneath them. They have created a work environment ruled by fear. People work in such conditions not because they choose to but because they may have little choice. It is a way to earn a crust and nothing more. Therefore, they will do the least they can get away with to satisfy their basic needs and to keep out of trouble but no more. Their loyalty is minimal because they resent the way they are treated.

Managers who have created or are sustaining such an environment believe from their own observations that their workforce are lazy, slothful and indolent and the only way they can get results is through threats, fear and carrot and stick methods. They are convinced that this is the only way to manage and will not accept that it is their behaviour and not that of the workforce that created the problem in the first place. They will not listen to any alternative no matter how convincing it might appear.

In great contrast, *Hoshin Kanri* demands a very different approach. The *Hoshin* organisation is seen as a community that has common objectives and in which each participant has a contribution to make. It is also founded on some fundamental beliefs about the nature of work. At the core of these beliefs is an understanding that everyone is the expert in his or her own job that people want to be listened to, perceived to be respected and to be developed to improve his or her achievements. Therefore, there are two key defining elements to *Hoshin Kanri*. On the one hand, it involves a very clear and well-defined structure of roles, responsibilities and metrics. On the other, it must concentrate on the development of its people with the aim to galvanise the creativity and job knowledge of its entire people to make it the best in its field for the pride and satisfaction of them all.

Pause for thought

Given that you have two alternative sources of supply. Which one might you trust the most? One in which people are ruled by fear or one where everyone is working towards their organisation being the best supplier?

Some autocratic managers are so entrenched in their beliefs that they probably appear incapable of change. However, it is surprising how readily some may become if their business is threatened. It is a way of maybe 'overcoming resistance to change. The business graveyard is positively filled with companies who are just too stubborn. It has been said that sometimes you need some Theory X to get some Theory Y!

SECTION 6: SUPPLIERS EXHIBITING NON-*HOSHIN* STYLE MANAGEMENT – THE WORST-CASE SCENARIO

The most striking features of such an organisation are the following:

No clear vision. Maybe there is a vision statement in the supplier's reception area but ask random employees what it means. The truth soon becomes clear!

No process ownership. Each function does what it believes to be its job then throws the product over the wall to the next department. There will be very little feed forward or feedback on requirements because people are 'supposed to know'! This is obvious in a blame culture. If people do not admit it, body language often tells the truth!

Most non-financial drivers of the business will be qualitative rather than quantitative and these will come in the form of vague slogans. There will be few if any tangible measures for anything other than those related to volume of work and those related to time constraints.

Not only will **competitor performance not be known**, in many cases people will not even know who the competitors are. This ignorance will often lead to crass complacency at all levels even when the threats should appear obvious.

A high proportion of **decisions will be based on opinion data** rather than factual. This will invariably lead to considerable sub optimization.

A blame culture will almost certainly prevail. As soon as anything goes wrong, which will be frequently, the question will always be: whose fault was it? Does the organisation practise small group activities such as quality circles? Does the work force appear motivated or do they appear to avoid eye contact?

The likelihood of insecurity at all levels of organisation is obvious. Such organisations usually suffer high levels of staff turnover, sickness and absenteeism.

Quick fix solutions are frequently not sustained when crisis management prevails.

HOW INFORMATION EXCHANGE BETWEEN AN ORGANISATION AND ITS SUPPLIERS CAN CONTRIBUTE TO IMPROVED OVERALL PERFORMANCE

One of the most serious disadvantages of 'multiple sourcing' is the resultant antagonism between competing suppliers. This leads to secrecy and deception. In contrast, with the 'Toyota Family' concept where suppliers are confident of secure long-term relationships, there is every advantage in cooperation. Consequently, it becomes possible for what might otherwise be competitors to share ideas, process improvement information not only with their mutual customer but with each other.

With some of the latest supply chain software, it then becomes possible for the original equipment manufacturer (OEM) to install this throughout the entire supply chain where not only can they inspect their supplier's inventory and work in progress, but the suppliers can inspect each other's as well. This enables the ultimate in just-in-time manufacture as illustrated in the case study on Toyota.

A further merit of this approach is the possibility provided for sharing assets. For example, where multiple sourcing takes place it might be necessary for each of the otherwise competing suppliers to have some product testing carried out using expensive equipment. Since unpredictability of further orders and that the equipment might not be required for much of the time it would be more economic to send this to a test house of some sort rather invest in such equipment which might be idle for long periods.

How providing managerial and technical support can help smaller suppliers achieve the corporate goals of their customers and contribute to their customers success

Philips, the Dutch electronics giant, was one of the first European companies to face up to Japanese competition in its market and survive in the early 1970s. Philips' competitors have long since faded from many people's memories especially in the field of domestic TVs. Some continue to this day for other forms of electronics and in the defence industry. Names such as McMichael, Sobell, HMV, Redefusion, Ultra, Decca, Pye and Ferguson mostly disappeared within one year of Japanese entry into the market.

Philips survived for two principal reasons. As soon as their market share began to drop they immediately conducted a defensive market survey to find out why the Japanese sets were preferred. There may have been many reasons, but one stood out above the rest. At that time, around 1974, there was a widely known statistic in the trade that warranty claims on a TV were typically approximately 6% per year. It was also known that the market was quite sensitive around that statistic. If someone launched a new model, and it was worse than 6% even by a fraction of a percent sales would drop rapidly. It was not the end user who was sensitive. The data was mostly collected by those in the TV rental market. They had to employ many service engineers to repair failed sets and as soon as one model proved worse than the rest, they quickly stopped promoting it. Philips found to their horror that the Japanese sets were only failing at the rate of 0.5% (later it fell to 0.1%)!

Now, to achieve 6% it was necessary to ensure that their parts suppliers could supply at something like 1% defective parts per lot (one defect per 100 parts). However, to achieve 0.5% overall set to set reliability, it was necessary for their suppliers to achieve parts per million defect levels with supplied parts. Neither Philips or their suppliers had any idea as to how to achieve such a goal. Philips then sent a letter to all their suppliers with the instruction. Within one year you must achieve defect levels of parts per million. If you do not, then we are dead and so in most cases will you.

Philips said we know that many of you are small companies and do not have the resources to find out how to do this and some of you might not even want to try. So, we are setting up a new policy. From the date of this letter we are setting up a new supply chain policy. All of those who are prepared to collaborate with us will be placed on a list of potential preferred suppliers. If you join that list, we will send our engineers to work with you on your production lines and find ways to achieve these levels of quality.

The policy worked and month by month the levels of performance improved sufficiently to enable Philips to remain in business. Most of their Western competitors made no such changes and the majority were gone in not much more than one year!

Assessing the role of suppliers in providing expertise to aid their customer's development and manufacturing processes

Hopefully, it can again be realised from the case example that openness and sharing is crucial to effective supply chain management. One critical aspect is the principle that a customer must have full knowledge of all aspects of the products and services with which they are being supplied.

Where this is not the case, it is almost certain the both products and the relevant operational processes will be sub-optimal. This is also true and, in some cases, particularly true where product testing has to be carried out both by the supplier and the customer. There are many cases where different standards are used, the calibration of test equipment is different with sometimes very costly consequences.

Evaluating the role of effective supplier selection in business performance achievement

The history of Western supplier/customer relationships is riddled with anecdotes in which all kinds of bad practice have been used to fool the customer into accepting products and materials that are not of the quality requested. Obviously, integrity is a critical requirement but with potentially new suppliers this cannot be taken for granted. Somehow, the customer must do some sort of research as to the likely performance of.

Analysing different criteria and methods for supplier selection and the types of industry or organisation where they may be used

Supplier monitoring can be carried out by second party surveillance, in which case the customer takes responsibility for evaluating the supplier himself.

Or he might decide to rely on some form of third party surveillance using something like the ISO 9001 Certification process. Of course, second and third party approvals are not mutually exclusive. It is quite possible that whilst a potential supplier might be ISO 9000 certified it must be remembered that this only applies to the integrity of the management system. It does not imply that the supplier is able to become integrated into a full just in time production process. If that is contemplated, then the customer will have no option but to carry out a full second party audit to ensure that the supplier's management system, production facilities, staff training and management style is fully compatible with its own processes and management system or can be adapted for that purpose and the supplier's management is willing to do it.

Alternatively, if we are buying certain goods, we may also either wish or be legally required to purchase products that have themselves been certified by some inspection standard. Again, this is also not mutually exclusive and can be an issue even where second and or third party schemes are to be used.

Two of these certification schemes are mentioned in Chapter 6.

- CE scheme for electrical goods
- BSI Kitemark scheme.

Internationally, there are also others.

In general, with the exception of the product certification schemes, which obviously only apply to the types of products that are relevant, all of the others are universally possible, and their selection will be largely based on feasibility, relevance and cost of using them or the risks of not using them.

SECTION 7: SUPPLIER RANKING ALSO KNOWN AS VENDOR RATING

Different organisations who use forms of preferred supplier status have a variety of self-developed rating systems. The following is the one developed by Toyota.

Supplier scoring methods used by Toyota:

- Grade 1. Supplier – high risk of total shutdown, can only operate with very high level of assistance
- Grade 2. Supplier – problem supplier – risk of causing assembly plant shut down
- Grade 3. Supplier – making progress but needs careful attention
- Grade 4. Supplier – good but not yet world class
- Grade 5. Supplier – exemplary performance and ability – long-term contracts.

Criteria typically met before agreeing to supplier self-inspection of products

Enabling the self-selection of products involves a high level of trust especially where the consequences of accepting a bad product is severe. Generally accepted criteria for this are included in BS6001 and it is advisable that this document be consulted. We cannot produce it here due to copyright issues, but the basics are as follows:

Incoming goods inspection

Irrespective of the confidence that has accrued in the ability of some suppliers to deliver consistently good products, there may in some cases even be legal or product certification reasons why it is necessary to continue with incoming goods inspection. Many of the schemes available include statistical sampling methods. These are particularly useful when the quality levels might not be high and are in the scope of sampling plans which are sensitive at levels of parts per hundred defectives or percentage defective per batch. British Standard 6001 is such a scheme but is flawed as explained in Chapter 3 in as much that it makes no mention of the economics of sampling, or the unreliability of the human or other means of inspection. If human inspection or sampling is used, the risks can be very high in some cases and much higher than most people realise. Therefore, the best possible selection of acceptable quality levels or the accurate calculation of sampling risk cannot be determined from the tables such as those included in BS 6001.

Analysing the merits of different supplier rating systems and explain the effectiveness of methods used to rank suppliers, including data on quality, price and delivery

The objectives of supplier rating are:

- To encourage suppliers to find ways to improve performance
- To identify suppliers who need assistance
- To develop a preferred supplier scheme
- To help move towards a just-in-time relationship.

There are two main methods of supplier rating:

1. Points method.
2. Cost ratio method.

Typical 'points' or weighting criteria given for supplier rating schemes are: Vendors or suppliers are given weighting according to their attainment of some level of performance, such as delivery, lead time, quality, price, or some combination of other variables. It may take the form of a hierarchical ranking from poor to excellent and whatever rankings the firm chooses to insert in between the two.

Typically:

1. Quality of suppliers weighting say, 30%.
2. On-time deliveries or meeting promises, say 25%.
3. Price, say 25%.
4. Other criteria, distance to location, financial status, or other relevant criteria, etc., say 20%.

For some organisations, it may come in the form of some sort of award system or as some preferred supplier status certification.

- To help the purchasing department make selections
- To provide the purchasing department with information helpful in subsequent purchases.
- Attempting to reduce subjectivity in supplier selection
- Providing criteria for discussion with suppliers
- Providing means for control of the supply base and KPIs for improvement purposes
- Enabling continuous improvement of supplier performance.

With the cost method, it is usually a simple case of lowest price supplier and taking a risk on the other possible issues. This can be acceptable in the case where the quality of the product or service is immediately verifiable on sight and that the product can be purchased 'off the shelf'.

Other considerations include:

- Developing a process for gathering information on product/service conformity for a multi-level supply chain
- Developing a process for reporting including measures that would be necessary if supplier quality no longer meets requirements

- Identifying and critiquing the methods available for monitoring incoming products and services.

Pause for thought

In your own organisation check how this is carried out if it is. Are you happy with the approach?

In Chapter 5 there is an extract from Jeffrey K. Liker's book *The Toyota Way*. For a benchmarking exercise, it would be worthwhile to compare these points with both yours and your key.

FICTITIOUS EXAMPLE FROM A DHIQC TRAINING COURSE

Cerif is an imaginary electronics company bidding for an important contract. The customer is intending to audit them to determine their capability but are conducting a pre-audit. They have sent their quality specialist John Osbourne along to look.

See what you can find that is not acceptable in a modern suppler partnership based on the principles and points listed above by reading the report by John Osbourne.

SECTION 8: PLANNING TO EVALUATE A POTENTIALLY NEW SUPPLIER

Supplier auditing – the 'Cerif' case study

Make notes based on what you have learned above. It will be a good test of your knowledge. Maybe you will think Cerif is quite good!

In this example, consider yourself to be either the character John Osbourne (you can substitute the name Sally Osbourne if you prefer!). John has gone to look at a potential new supplier prior to planning a full second party audit. Cerif manufacture a video movement detection and character recognition device that John Osbourne's company, a Japanese subsidiary of a large multinational, plan to build into their own security product to reduce theft in car parks.

Once they have selected their suppliers they intend to provide them with training and a period in which to gain competence prior to full scale production. A study shows that, currently, their patents are well protected and that there is a huge international potential market.

You are familiar with ISO 9000 (the latest version), the Toyota 14 points and both Deming and Crosby's points but you are mainly concerned with the Toyota approach.

1. John Osbourne is a quality professional who has successfully completed the DHI Diploma in Quality Leadership and has been invited to meet Peter Talbot, the quality manager of Cerif Limited (a small company located west of London, UK), at 0900 that day to determine whether they are a fit company to be considered suitable for their preferred supplier network and to operate a just-in-time supply chain policy based on Toyota's.

2. In an earlier telephone conversation, Peter had told John that the company's management system was based on total quality principles, which had been introduced by another firm of consultants some five years earlier. Therefore, Peter believed that they would not have much difficulty in satisfying the requirements of ISO 9000:2015 at least.

3. John arrived 15 minutes early by car; the map that Peter had sent by fax was adequate but missed two new roundabouts that had been installed since the map was printed. Luckily, John knew the area.

4. John pulled up at the security gate. At the entrance was a large sign with the words emblazoned 'Our Customers Are No 1'. A good start, thought John. He wound down his window and offered the invitation from Peter to the security guard. He was then directed to the rear of the plant to the visitors' car park. It had started to rain. He drove passed reception, then saw signs stating directors' parking adjacent to the reception before reaching the visitors' car park, which was just rough ground to the rear.

5. Soon, he was in the reception area. It had recently decorated with comfortable chairs and some magazines spread neatly over a table. He noticed a number of certificates on the wall including a certificate for ISO 9000:2005 and a large slogan that said: 'VISION – It is the policy of Cerif Limited to be the unchallenged world leader in microcomputers and video terminal products. This will be achieved through the knowledge, skill, creativity and job knowledge, highly motivated workforce, customer focus and a lean and efficient organisation'.

6. The receptionist was a lady probably in her mid-30s and smartly dressed. John asked her to contact Peter Talbot. 'Is he expecting you?' she enquired. 'Yes', said John and showed her the letter. She made a quick call and then said that Peter would be there shortly, would he care to sit down?

   'Fine. By the way, I was looking at the mission statement on the wall. What does it mean to you?'

   'Not much. It was there when I joined the company, but I have only been here for six months.'

7. John picked up a newsletter. It contained mostly articles about recent acquisitions, the launch of a new product or two, plus a picture of someone who had recently retired after 30 years of service and some explanation of management changes at the top. Apparently, the previous managing director, Tom Chadwick, and his co-director, Mal Shaw, had been replaced at short notice by Vernon Wagstaff as managing director and Wayne Howes as finance director. There was also a letter from Vernon Wagstaff explaining his ideas about bringing the company into the twenty-first century and winning back markets.

8. Peter arrived. He suggested that due to a small problem that had unexpectedly arisen, the quality department might not be appropriate for their meeting, so he had booked a conference room.

9. It was intended that they should meet Vernon Wagstaff, the managing director, for lunch, would that be OK?

10. On the wall of the conference room, John noticed a copy of the same slogan he had seen in the reception area. In fact, he had also noticed it in several of the offices they passed along the way.

11. 'I see that you have a vision statement.'

    Peter looked at it for a moment and made a wry smile. 'Yes. It was the managing director's idea when he returned from a conference a year ago.'

    'Did he write it himself?' John inquired.

    'I believe so, he seemed very pleased with it when it was first put up. It probably came from his last company.'

12. 'Is it ever referred to?' asked John.

    'Usually when something has gone wrong there are one or two cynical references to it but apart from that, not to my knowledge. To be fair, it may have more meaning to the directors.'

    'Has it ever been explained to anyone?' asked John.

    'Not really, but I guess it is self-explanatory.'

13. Peter went on to describe the company, its products and markets. There were 450 employees at that site with sales divisions in Germany, Singapore and Italy, all about equal size, and an expected acquisition in the USA that was causing contractual problems. All of them had ISO 9000:2005. Some models of the product were designed on this site, which was the headquarters of the microcomputers and video terminals division of the company, and other products were bespoke to suit customers' needs. The company did not think they would benefit from the latest ISO 9001 due to the cost, so decided to keep things simple.

14. 'You told me that you had a total quality programme, can you describe it to me?' asked John.

15. 'Sure, we began four years ago just after a serious financial crisis in which we had to remove 50% of the labour force, it was terrible, we lost £36 M in that year and we nearly went down.'

    'What was the problem?' asked John.

    'Our biggest customer pulled out on us and they had been taking more than 70% of our product.'

    'Why did they pull out?'

    'There is a long history to it, late deliveries mainly, but the final straw was due to our CE mark. We had overlooked the date when the product we were selling them should have been certified but we made out they were OK. Unfortunately, they failed a radiation emission test and the truth came out.'

    'What happened then?'

    'The chief designer was sacked but apart from that nothing. They were too busy with the redundancies.'

16. 'Could it happen again?'

'I hope not, but it is hard to tell. Pressure to get new designs into production is our biggest problem. Unfortunately, the old managing director was a sales man originally. It will be interesting to see what Vernon Wagstaff turns out like. Supposed to be a Harvard whiz kid. Nice guy but a bit impatient.'

17. 'Then we were bought by Global Holdings who insisted that we had to introduce TQ. Were the directors supportive? At first they were, but they do not get involved these days.'

'How is it supported then?'

'Tom West is the TQ coordinator. He is very enthusiastic and a good guy but doesn't get much help from anyone. It works quite well in the manufacturing department because the manager there worked for a company that had TQ and he swears by it but with the overtime and other things they do not get much time to do anything.'

18. 'Can we meet Tom?' Yes, I think he is in this morning. Good, I would also like to meet the manufacturing manager and some of the directors if possible, said John.

19. Tom's office was at the rear of the factory. On the way there, John looked around him. He noticed that there seemed to be a great deal of inventory between the processes. About 40% of the plant was running with several of the machines being set up or idle. The incoming goods department took up about 20% of the available floor space and it was about 80% full. Finished goods stocks were about the same.

20. John noticed several quality slogans around the walls including a few copies of the VISION statement. One of them said 'quality makes customers, customers make jobs, do it right first time'. John also noticed an LED display board that said 'This week's production target is 750 units'.

21. 'Do you usually meet the targets?' asked John.

22. 'Mostly. We operate a payment by results scheme, so the workers will always try to make as many products as possible.'

'How does that affect quality?' asked John.

'Well, obviously, the workers will try to hide defects when they can, but we probably have one of the most effective inspectors in the business. He was one of our highest earning process operatives himself once, so he knows all the tricks and not much gets past him. Also, anyone found trying to conceal bad work will have it deducted from his wages and be reprimanded with a written warning, so they know they stand a chance of being sacked if they keep doing it. It is a good system.'

'Are the workers happy being treated like this?'

'I am not sure, I have never asked them, but the wages are good so very few leave. Also, we have a good canteen and a sports club that some of them like.'

23. How about absenteeism and sickness? 'That is a bit worrying and slightly higher than the local average.'

24. 'Industrial accidents?'

'We are running at something like 850-man-hours per lost time accident right now according to our health and safety advisor.'

'Is that good or bad?'

'We are not sure, but it sounds good to me.'John made a note to check.

25. 'How are new workers trained?' asked John.

'We used to have an industrial engineer, Fred Willett, who broke all the tasks down into very small units. This effectively deskilled most of the work, so training is easy. When a new employee joins the company, we sit them next to one of the better skilled operators for a few days who shows them what to do.'

26. John noticed an SPC control chart on a board near to one of the machines. It was well laid out and the marks on the chart were up to date and there were several sheets underneath going back several weeks. It seemed from the data that the process was very well in control. All the dots on the chart were close to the mean with none being anywhere near to be a 'single standard deviation' in spread.

27. John asked 'Do you know the number of defectives on this process?'

'Yes, it is around 5%.'It seemed odd to John that the dots were so far from the action lines and yet the process was clearly quite variable. He made a mental note to look deeper into this. He had ideas about the possible reasons and how he might test his theories.

28. 'Who fills in the charts?' asked John.'One of my quality technicians, he learned to do it at night school.'

29. Tom West was a casually dressed man in his mid-thirties. His office was small but neat with a clear desk.

30. Tom explained his job and his experience. He had not been trained in TQ at Cerif but had learned it whilst studying for an MBA during a previous employment. Yes, he was aware of tools such as the Fishbone diagram, Pareto, and process mapping and did use them when he could. They usually had three or four improvement teams at any one time depending upon the number of problems they had.

31. John asked how the problems were identified. Usually, the manufacturing manager would ask Tom to set up a team if there was a problem of late delivery or scrap or some supplier problems.

32. Other managers could ask for teams to look at their problems but that was rare.

33. John asked what training the teams were given. Tom said that it depended on the problem. Obviously, the teams were made up of people with knowledge of the process and familiar with the situation, so they did not need much training.

'What about problem-solving tools?' asked John.

'If they need it, I give them some help with brainstorming or the fishbone diagram, but we do not find them useful for much of the time.'

34. Later, John had lunch with the managing director. At first, he did not appear to appreciate the consultant's intrusion into his territory and said so. Apparently, he had some bad experiences with consultants in the past and assumed that they were all the same. This included the consultant who had first introduced TQ.

35. John did not react. He enquired about the vision statement he had seen everywhere.

   'I created it', said the managing director, 'to try to develop a spirit of quality mindedness in the company.'

   'Did it work?' asked John.'I think it did for a while.'

36. 'What measures did you use to decide whether it worked?' asked John.

   'Mostly the opinions of the other four directors. One or two of them were convinced that it had made an impact, but the others did not agree.'

   'Were there any tangible measures to determine success?' asked John.

   'What sort of measures do you mean?'

37. 'Well, let us take a look at the statement again.'

38. 'VISION – It is the policy of Cerif Limited to be the unchallenged world leader in microcomputers and video terminal products. This will be achieved through the knowledge, skill, creativity and job knowledge, highly motivated workforce, customer focus and a lean and efficient organisation.'

39. 'It says, "unchallenged world leader". It sounds good but what does it mean? How can you tell if you are the leader or not? What are the success criteria?'

40. 'I suppose "market share".'

   'Fine, but do you know your market share? Do you know the market share of your competitors? Do you know if they are increasing their share or if it is decreasing? Do you know if there are others who might enter the field?'

41. 'Of course, market share is not only determined by the competition, the customers, users, distribution networks and others also have an impact. Do you know what they think of you compared with the competition? How would you find out?'

42. 'We think we know the answers to some of those questions but not all of them. What does all of this have to do with our getting the contract for your assemblies?'

   'Well, we will need to talk about that after I have prepared my initial report, which I will share with you before I leave. It is our hope that you will find all of this helpful because we have concentrated in recent years in trying to be the best in class in our field. We have also learned that unless our suppliers adopt a similar approach to this as ourselves, we cannot achieve our goals. Therefore, this audit is not a pass or fail examination, far from it. We expected there to be issues. What will determine whether we end up working together will be how you react to the suggestions in our report. We believe that what we have to say will be beneficial to you whether you work with us or not. If you wish to adopt the measure that we will suggest, then it will be in our interest to help you to achieve the requirements.'

43. 'If you did know the answers to my questions and you found that your performance was not as good as you would like, what would you do?''Ah', said the managing director, 'we are probably better there, we have the TQ programme, you see.'

44. 'Can you tell me how it works?'

   'Yes, following our monthly board meetings we identify quality objectives for each of the managers, this is something that we learned from ISO 9000:2000. We also evaluate the improvements made by the managers since the last meeting and give them a hard time if the results are not achieved.'

45. 'Thank you', said John. 'Do you praise them when they have made achievements?'

'Sometimes, it depends what it is. If it saved a lot of money, we might make a cash award to the individual who thought of it.'

'How about when the idea came from a team?'

'Yes, we would do that too.'

46. 'Can you tell me why you do not have the current version of ISO 9000 even though you clearly design your own products?'

47. 'We are in a very competitive market and innovation is our lifeblood. We must be the first in with a new design otherwise the high-value contracts will go to our competitors.'

48. 'Why does that stop you going for ISO 9000?'

'Sometimes, we have to short circuit the design process that would be too long winded if we had to comply with those requirements.'

'Thank you', said John, 'you have been very helpful.'

49. Later, John said to Peter, 'I have seen all that I need to see before planning my detailed audit, which I would like to do as soon as possible. Before I leave can I please come to the quality department and see the problem that concerned you this morning?'

50. The quality department comprised a product test area for both incoming goods from suppliers and an assortment of automatic test equipment used for checking their own production. There was also a large repair shop and a small standards room. In the centre of the room was a large pile of boxes of finished products.

51. 'This is what came back this morning', said Peter.

'What is the problem?' asked John.

'The first batch we sent the customer was made in the prototype shop', said Peter.

'Unfortunately, John Wadham, the sales director, had seen the design at the conceptual stage and had promised International Gulf Products that they could have the first batch last January. The only way we could do that was to sell them the prototypes. That was OK but when we went into full production later, the display on the LEDs was different because we could not get any more of the originals. Also, the transmission range is wrong.'

'Why is that?' asked John. 'Probably it is because the suppliers have used the wrong chip. The drawing shows a specific component, but we allow them to use alternatives with equivalent performance characteristics but do not always state what they are.'

'Surely', John said, 'that is bad practice?'

52. 'I agree', said Peter 'but sometimes the correct items are unobtainable, and we would not achieve delivery dates if we did not allow it. Also, there is sometimes a discrepancy between the results the suppliers ATE and ours.'

'Is there a calibration problem?' 'Sometimes, but, unfortunately, the equipment they have is on loan from us. So, they say that they are testing the product to our spec.'

53. John looked at his watch. It was time to go. On the way home, he thought about what he had seen.

54. He was now thinking about how he might plan the detailed audit.

Benchmarking thoughts

- Supposing that you had been in the position of John Osbourne.
- What did he know already from what he had seen?
- Identify what you could see that was seriously deficient in their approach to business from a supplier point of view.
- What were the questions that needed answers?
- He needed to construct an audit plan. Who should he see and what should he look for.
- You can refer to the Toyota executive brief for principles and compare them with John's experience at Cerif.

On the basis of what you already know about sound business management systems

1. Brainstorm the deficiencies that you think you have found already.
2. What do you think is good?
3. What now looks highly suspect and needs to be looked into?

Whilst we acknowledge that the legal aspects of supply chain management are normally the responsibility of the purchasing department, there is a very thin line between these responsibilities and those of the quality function. It is therefore important, in our view, that those concerned with supply chain management should at least have an appreciation of the most important aspects. Therefore, we have included them here.

SECTION 9: THE LEGAL ASPECTS OF SUPPLY CHAIN MANAGEMENT

We only intend to include some of the key highpoints here. For a more in-depth treatment, it is strongly recommended that the principles included in Chapter 6 on legal and regulatory aspects is considered. Bear in mind also that in the case of product liability lawsuits, the injured party can act against the entire supply chain and then focus on the one most likely to be vulnerable.

Partnering arrangements

Where items or materials are purchased on a regular basis, a customer may decide to develop a partnering arrangement to contract out the associated risks to the supplier.

The customer will make a commitment to the supplier and, in return, the supplier will take on some or all the risks associated with late or non-compliant delivery. The customer may categorise items of equipment and materials as well as specify limits for financial penalties.

Alternatively, in the case of products liability risks where the customer has large resources, it is common for the customer to underwrite or take over the risks of its supplier. This is particularly the case in the automotive industry where product recall is a high-level risk with a high rate of occurrence.

It also carries with it the threat of potential bankruptcy if products liability lawsuits are successful. The reason for this is that in law, when a products liability case arises, it is in the interest of the pursuant to sue everyone in the supply chain – initially from the retailer back, and then focus on the one most likely to be able to pay. Many suppliers to the large automotive companies would not trade if they were confronted with this level of risk. Partnering may also reduce the risk of unpredicted bankruptcy of a critical supplier.

Insurance

Where the use of partnering arrangements is impractical then the customer may identify those risks to be insured in-house and on the open market.

Risk-tolerant work packages

To reduce the cost of insurance, either in-house or by a third party, the customer may develop risk-tolerant work packages. The use of suppliers' standard items that have already been extensively tested in previous use, rather than custom-built untried items of equipment, is an example of this type of package. Boeing, for example, has a policy that each new design of aircraft must contain at least 70% components in use in earlier models. The adage 'innovate at your peril' is extremely apposite.

Prototype and product testing

To reduce the risk of recall and prosecution under the product liability laws, prototype and product testing and validation of service has been used as an effective form of risk control. There are two aspects to this approach:

- Risk prediction – this can be done using such tools as FMEA and fault tree analysis. These tools are quite like each other and involve such techniques as brainstorming, affinity diagrams and relationship diagraphs to predict high likelihood risks and possible countermeasures. In the case of management and financial risk there is an adaptation of FMEA known as a process decision programme chart. It can be very effectively used for the prediction of management-related problems, identification of countermeasures, identification of problems caused by the countermeasures on an iterative basis until either achievement of the desired outcome can be reasonably assured or the mission abandoned
- Risk analysis – this involves life testing, accelerated life testing, environmental testing, Weibull and other forms of failure analysis. From the results of these tests preventive strategies can be developed. For example, it is known

from the results of extensive product testing that sodium streetlights run for a predictable 8000 hours. Rather than replace each one individually on failure, it is more economic to replace all the lights in the area whether they have failed or not. This is known as a block replacement policy and is extremely economical.

Risk reduction resources

Risk reduction resources are generally provided in the form of inspection, expediting or auditing services, but may also include risk management planning resources. The extent to which inspection and the associated expediting activities are undertaken for any item of procured equipment, will depend on the perceived risks of items being delivered to the site or the receiving warehouse – not in accordance with the specification requirements and programme delivery requirements. It is therefore industry practice to carry out a risk assessment that addresses the consequences, in terms of cost, for either of these events occurring, i.e. in non-conformance to quality requirements and late delivery.

Depending on the industry sector, this assessment will be carried out for all or only the major categories of procured equipment. A typical risk assessment will address the following topics:

- Criticality of the product based on the consequences of failure. A criticality rating will typically address the following factors:
  - Product safety implications
  - Design complexity and maturity
  - Complexity of production process
  - Product characteristics
  - Environment issues
  - Operation issues
  - Direct and indirect economic consequences of failure
  - Cost of rectifying defective items, if any
  - Cost from consequential delay to programme in rectifying defective items
  - Cost from delay to programme for late delivery of items
  - Cost of inspection and expediting as a percentage of purchase price
  - Whether a vendor quality plan is required.

Once these costs have been assessed, it is necessary to identify the following:

- Pre-award audit of the supplier's QMS
- Pre-inspection meeting
- Intermediate inspection visits, if needed, how many and for what
- Final inspection visits
- Expediting visits, if needed, how many and for what aim.

Risk assessment techniques

It is not the intention of this book to cover the various tools and techniques currently in use to manage and assess risk – hazard and operability studies (HAZOP) and

FMEA to mention just two. These techniques have been covered in great detail elsewhere. The only exception is risk-based decision-making, which has received less exposure and is included as an example of one of these techniques.

Most of the decisions made in the working environment will benefit from an analysis of the risks involved. In practice, there is usually too little time, and the actual process of risk analysis is perceived to be too lengthy for this course of action to be undertaken with the associated record being placed on file.

There is a clear need for a simplified system of identifying and analysing risk, so the process can be applied on a more regular basis. Those decisions classified as minor, but that can still materially affect the prosperity of a company or the success of a project, will benefit. Paradoxically, however, it can also be argued that without risk analysis, an organisation remains vulnerable to all levels of risk and will be the least prepared to deal with them if any materialise. The process of risk-based decision-making is suited to these minor decisions and, under normal circumstances, it can be conducted in about one and a half hours, usually involving two or three people. The result is that for the investment of approximately three hours of time spent on analysis, a decision can be taken having considered and budgeted all risks involved and the related outcomes. Risk-based decision-making can also be applied to major decisions such as the selection of joint venture or alliance partners, but the time required to develop the analysis will be even greater. Risk-based decision-making is typically used for the following types of decisions:

- Review of invitations to tender
- Selection contract agreement type
- Placing of subcontracts
- Equipment selection (the design process) and procurement
- Resolution of problems encountered on projects
- Development of plant maintenance philosophies
- Selection and employment of company personnel.

It can be seen from the sections in this chapter that supply chain management is an area that lends itself to considerable creativity. The Toyota system is impressive but so are many others. What is important is to move away from multiple sourcing and the related adversarial relationships approaches that are still very common and to develop partnering relationships both down through the supply chain hierarchy and to encourage them to collaborate across the chain as well. Toyota refer to their own version of this as the 'Toyota Family'.

Try to encourage suppliers to operate *Hoshin Kanri* and if they do not know how, then help them on the basis that it is in your interest that they do so. Reward the best with preferred supplier status. It will make them feel good and will give even better service and it will help them to pass the benefits back to you through better service.

Overcoming resistance to change

Frequently, supply change management involves the requirement that a supplier or potential supplier should make some quite dramatic changes to his management system and approach before we are prepared to enter any sort of relationship. Such organisations can react in many ways. In some cases, they will decide that they would rather turn down your business rather than make the changes. This will often be the case when the customer is small, and the supplier is large or when only small quantities of product are required. We offer no solution to this other than either to reluctantly accept the situation, make the products yourself or find yet more possible sources.

In other cases, the supplier might be eager to win the business and to make radical changes. However, they might have no idea as to how to go about it. This will not only present a problem at the top level, it will ripple all the way down as the changes will, in the end, affect everybody. In this case, it is as well to understand the mechanics of change and resistance to change. Here, Dr J. M. Juran has been very helpful with his video 'Overcoming Resistance to Change'. This can be located on YouTube and is recommended.

SECTION 10: SUMMARY OF THE KEY POINTS OF THIS CHAPTER

Most of companies today outsource more than 60% of the work content of their products and services and the trend is towards higher and higher levels with the possible consequence of greater supplier dependency. However, it is now becoming widely accepted, at least amongst the pacesetting companies, that the supply chain should be regarded as an extension of our own processes and responsibility for their performance is not delegable. Whenever an organisation devolves responsibility to a supplier, they are losing control of the process and at some time this will prove painful.

The current situation in many countries of the world

Multi-sourcing or single-sourcing?

Prior to the 1990s – many organisations used multiple sourcing and adversarial relationships to obtain their requirements from suppliers. There was widespread use of vendor rating schemes. This system has not gone away, unfortunately, much as some would like to believe.

Vendors were penalised either by awarding future contracts to competitors therefore starving them of expected business or strengthening their competitors in the short term. They also penalised them by refusing to pay, with the excuse that they had not received the goods they had contracted for. This produces serious cash flow problems for the supplier and sometimes even puts them out of business. This is bad news for the supplier and his employees, but it could be short sighted on the part of the customer who, therefore, has one less source to choose from.

Just-in-time supplies

1980s–1990s – the majority of companies focussed on reengineering supply chain cost structures by demanding just-in-time supplies and running stock levels down to as close to zero as possible. This is claimed to have reduced costs by some 33%. However, it was done very badly. Suppliers were not instructed how they might be able to do that using systems such as Kanban and other so-called lean techniques with the consequence that the excess stocks were still held in the system but by the supplier rather than the customer. All it meant was that all the previous bad practices remained, and the stock simply moved from one step in the process back to the former. The cost was still there!

2000 plus – it was claimed that quality no longer provides an advantage, only a qualifying factor. This was the erroneous belief that companies understood 'quality'. This may have been true for a few but, in our opinion, very few and this is true to this day. If anything, quality knowledge has regressed in the past decade rather than progressed – consequently, the focus is now on revenue growth through customer satisfaction but even that is still the topic of heated debate as is shown in the quality-related forums on the internet. The costs will still be in the supply chain!

The impact of Brexit! Whether or not the UK is either in or out of the EU. One thing that the debate has thrown up is the importance and intricacies of Just in Time supply to a nations economy. It has taken the UK and the EU some 30 years to catch up with Japan's superiority in this respect, but it can be destroyed almost overnight.

One-real change

With the enablement of new technologies such as the internet, outsourcing and partnering with other enterprises are becoming more commonplace as companies seek to share the burden of demand for more complex products and more responsive services. The building blocks of successful supply chains are numerous, and their interactions are complex.

Pause for thought

If multi-sourcing results in adversarial relationships, devious behaviour and poor quality, what might be the risks associated with totally single sourcing?

If quality-related costs are, as is claimed in other chapters, some 20% or more of our sales revenue, the probability is that this is equally true of our suppliers. **This cost will be passed on to us**. If we are competing with others who are not only aware of this but who treat their suppliers as extensions of their own processes, then they will be reducing these costs significantly year on year and we will be left behind!

Summary of learning

What you will have learned in this chapter

- Advantages and limitations to customer and supplier of multi-sourcing short-term supplier relationships compared with near single-sourcing long-term supplier relationships

---

- The risks concerned with disruption of the supply chain
- Overview of the respective roles of the quality and purchasing functions
- Trends in supply chain management due to globalisation and computerisation
- Supplier monitoring
- Comparison of vendor rating schemes, the use of 'preferred supplier status and other supplier motivation schemes such as 'supplier days' joint advertising, etc.
- Incoming goods control, supplier traceability
- Plan a pre-audit of a potential supplier based on the fictitious 'Cerif' case study.

9 *Where to go from here?*

If you have just read this book from cover to cover without pausing for thought and then acting on it, possibly you might have finished it in, say, a couple of weeks. However, since most of the content is included in our Diploma course, it is normal for someone to take between two and three years of serious study to reach the end. The book contains a lot of content but please bear in mind that for much of the content I have only skimmed the surface. Hopefully, I have gone deep enough for it to have meaning but also to stimulate interest in drilling deeper.

One additional thing I would recommend is to look at the recommended reading list. Also, the Juran Institute have very kindly posted many of Dr Juran's old videos on YouTube and also their video product 'Quality Minutes' – they may be old, but the content is as relevant now as it was when it was created. There are two videos of importance. One is on the subject of 'overcoming resistance to change' and the other 'operator controllable errors'. There is also a video, not in the same series, entitled 'A message to upper management', also by Juran. There is a 45-minute and a 10-minute version. The 10-minute version is fine and makes the point.

If you have made a lot of notes in your notebook related to this text and the 'Pauses for thought' we will also assume that you want to make changes in your organisation. If you happen to be the chief executive, you do have certain advantages over everyone else. If people disagree with you then generally, they should have a pretty good argument! Hopefully, the content of this book will have given you the ammunition that you need to make the changes you choose.

If you are not the CEO, then maybe you have selling to do. The ammunition to help you do this was deliberately included in Chapter 1 where the causes of high-quality cost were highlighted. Also, if you carried out a project whilst studying Chapter 4 then hopefully you might use that as an in-house example of the potential of this approach. Again, hopefully the book will help but if not and you are keen enough to keep trying, there are other options.

One key question might be, where do we start? This advice is addressed to the CEO. If you are other than this then you must address this to him/her.

Step 1

Using the notes that you made, first of all using your organisation chart, highlight the issues that must be addressed, and the level of management involved. For example, it may be that you need to establish a steering committee. We would recommend that this is Step 1 anyway.

Step 2

Appoint a facilitator to work both with the top team, the steering committee and then all of the remaining parts of the programme.

The selection of a good facilitator is crucial to success. They need to have special personal qualities as well as competency. These are as follows, not in any order:

- Enthusiasm. It is important that they want to be facilitators. For the individual concerned this is a serious point. For most, it is a career change. They will have some questions. How serious is the organisation? Is it a whim that might pass or can they convince you that it is a permanent change? What if it is a whim? Can I get back on track with what I had planned for myself?
- It needs to be somebody who, when appointed, is likely to want to live, sleep, eat and dream this challenge, to contact others doing the same and to network and exchange experiences.
- It must be somebody who is liked, trusted and respected at all levels and is approachable. Others must think 'they must be serious about this or they would not have appointed him/her to manage the implementation'.
- They must be knowledgeable in the topics they are to introduce but, initially, not necessarily experienced. It would be nice if they were, but such people are not easy to find.
- To 'facilitate' means to make easy. The facilitator does not actually do anything as such. Their skill is in helping others to make their own plans, to help implement them and to keep an eye on the results. They should be concerned with building people, building teams and giving confidence.

Also, I have mentioned a few organisations that have already implemented these concepts, in some cases decades ago. Check them out. The advice they can offer will be priceless.

JUSE in Japan, hold an international seminar on TQM in early July and early December every year. They have delegates from some 30 or so countries every time. The trip includes visits to some companies but if you take a large enough group, they will often lay on some custom-made visits to suit your needs. If this does not work nothing will.

Step 3

It is recommended that the next step is with the help of the facilitator, the steering committee might revisit the corporate vision statement if one exists. If not, then you should create one based on the advice given in Chapter 5 in the explanation of *Hoshin Kanri* and follow the implementation process explained there. It is also advisable to use the David Hutchins book *Hoshin Kanri A Strategic Approach to Continuous Improvement*, as it contains far more detail. From the 'vision', the implementation process will help you cascade down through the organisation. As you go, you will find that some of what is important is already in place. This is a bonus. If not, then you will have highlighted the work to be done and the facilitator can help with that.

One note of warning: make sure that you do not overload the facilitator. It is easily done. Often, it will be a part-time appointment. If it is, they will have a boss. Previously to this, the boss was accustomed to having the facilitator full time and has certain expectations. This will result in a conflict of interest and the build-up of considerable pressure in the facilitator. They may not complain but either one or the other of the jobs will suffer.

The appointment of a full-time facilitator is best, but still the work will increase as the programme matures. It needs to be kept under surveillance especially if the programme included the introduction of quality circles or any form of self-directing work teams.

IMPLEMENTATION PLAN

The use of a Gantt chart is a great way to do this. Now look at the gaps and you need a strategy to start filling them. The competency wheel in Figure 9.1 might help you. The following model is adapted from a paper that Mike Hutchins created for DHIQC some time back entitled 'Quality leadership through a new lens'.

Quality leadership through a new lens

1 The challenge

The quality profession is at a watershed. The role of quality professionals has never been so blurred. Change is unprecedented. How ready are you to embrace the future? What competencies are needed to succeed in this new, challenging environment? The whole way business is conducted is changing. It is not just quality that is being impacted. Finance, HR, IT and every other function are facing similar challenges.

Table 9.1 summarises some of the changes facing all organisations and the quality profession:

Table 9.1

Past	Present and future
Centralised quality department	Lead improvement efforts. Quality integrated into business systems and culture
Local, vertical control of supply chain	Worldwide sourcing
Product quality seen as a dierentiator	Perfect product quality assumed
Focus on control and defect reduction	Re-invention, reduced cycle times
Predictable, loyal customers	Worldwide customer engagement

Key questions

1. What competencies are needed in the future for the quality professional?
2. Where do you need to grow personally to become a leader in quality?
3. What training and career development path is required for an aspirant quality leader?

This outlines a future vision of the key attributes of a quality leader, through the lens of three 'archetypes' – the 'steward', the 'strategist' and the 'leader'. To be truly effective in any organisation, strengths in all three areas is required.

2 Quality leadership competencies

The competency wheel depicted below (Figure 9.1) is a blueprint for anyone involved in facilitating the integration of the quality sciences and disciplines into an organisation and aspiring to be in a quality leadership position.

The wheel starts with the quality leader, or 'chief quality officer' (CQO). As you peel back each layer, there is more detail around what competencies are required to be successful in that particular archetype. So, for example we see the quality strategist requires expertise in 'strategy', made up of 'systems thinking', 'organisational design' and *Hoshin Kanri*. Below is a description of the competencies within *Hoshin Kanri*, or strategy deployment:

Table 9.2

Competency	Description
Hoshin Kanri	1. Develop mechanisms for translating strategic goals into meaningful, measurable operational metrics.
	2. Create a culture of continuous performance improvement and a system of feedback loops throughout the organisation (nested PDCA)
	(Re-defining quality. White paper by Mike Hutchins, p. 3)

DHi Competency Wheel™

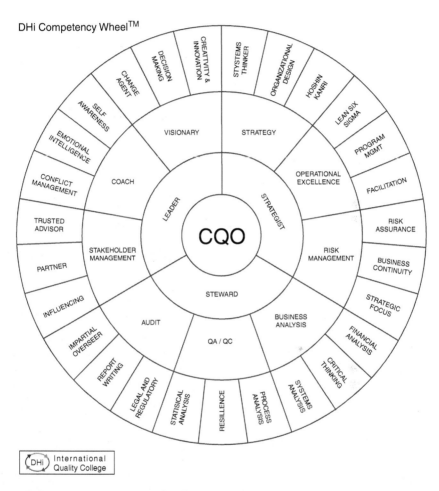

International
Quality College

Figure 9.1   Competency Wheel

The three archetypes

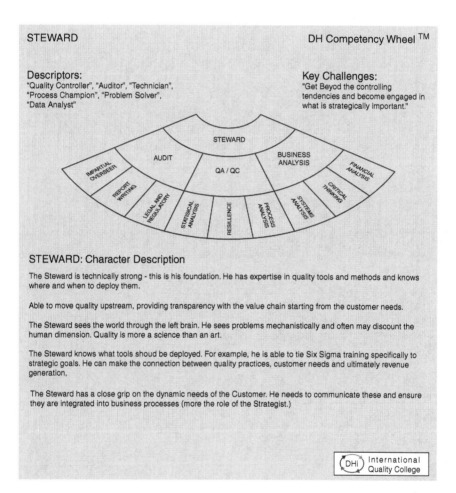

STEWARD   DH Competency Wheel ™

Descriptors:
"Quality Controller", "Auditor", "Technician",
"Process Champion", "Problem Solver",
"Data Analyst"

Key Challenges:
"Get Beyod the controlling
tendencies and become engaged in
what is strategically important."

STEWARD

AUDIT

BUSINESS
ANALYSIS

QA / QC

IMPARTIAL
OVERSEER

FINANCIAL
ANALYSIS

REPORT
WRITING

CRITICAL
THINKING

LEGAL AND
REGULATORY

STATISICAL
ANALYSIS

RESILIENCE

PROCESS
ANALYSIS

SYSTEMS
ANALYSIS

STEWARD: Character Description

The Steward is technically strong - this is his foundation. He has expertise in quality tools and methods and knows where and when to deploy them.

Able to move quality upstream, providing transparency with the value chain starting from the customer needs.

The Steward sees the world through the left brain. He sees problems mechanistically and often may discount the human dimension. Quality is more a science than an art.

The Steward knows what tools shoud be deployed. For example, he is able to tie Six Sigma training specifically to strategic goals. He can make the connection between quality practices, customer needs and ultimately revenue generation,

The Steward has a close grip on the dynamic needs of the Customer. He needs to communicate these and ensure they are integrated into business processes (more the role of the Strategist.)

DHi International
Quality College

Figure 9.2   The steward

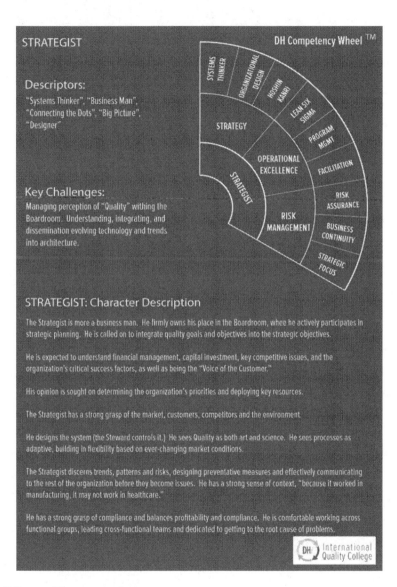

Figure 9.3   The strategist

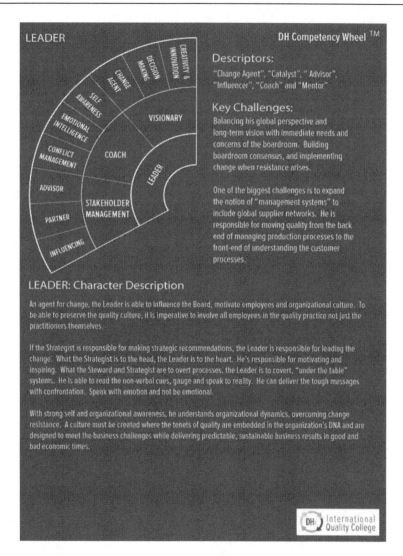

Figure 9.4   The leader

4 Learning and development

Individuals

- Study for the DHIQC Higher Diploma in quality leadership. All of the steward competencies are covered here.
- A strong focus on technical skills, tested through assignments, ensures a solid grasp of the theory and application of quality tools and methods.
- Read more about the Competencies in the Competency Wheel (www. hutchins.co.uk).

- Enrol for the DHIQC course on Facilitating, Coaching and Mentoring for Quality.
- Complete a leadership assessment tool such as the Leadership Circle Profile, Emotional

Intelligence 2.0 (http://psychology.about.com/library/quiz/bl_eq_quiz.htm) or the Trusted Advisor.

- Assessment (free: http://trustsuite.trustedadvisor.com). Identify your strengths and weaknesses (simple assessments with the book *Strengths Finder*).
- Find yourself a coach or mentor to work with you in developing your career (contact DHIQC).
- Request you attend a leadership programme within your organisation.
- Spend more time outside your functional area and comfort zone, request internal transfers, secondments and different projects.

Organisations

- Build mechanisms within your talent management process to identify emerging quality leaders.
- Develop internal leadership programs, or send emerging leaders on such programs.
- Incorporate 360 leadership and quality assessment tools into performance appraisals.
- Formalise a coaching and mentoring programme.
- Incentivise staff rotation programmes.

5 Conclusions

Through use of these three archetypes, the steward, strategist and leader, my intention is for you to start reflecting on where you are. What are your natural strengths? What is your default style? What areas do you see yourself needing to grow? By redefining what it means to be a quality leader, accepting the challenging new realities that are facing the quality function, so you have to target to aim towards.

6 About the author of this section

Mike Hutchins is a CPA/chartered accountant with 20 years' international consulting experience. His expertise is in managing large global change programs for clients such as Vodafone, Cemex and Wolseley. He has worked in multiple industries, including telecoms, automotive, US Federal Government contracting, logistics and distribution, manufacturing, retail, healthcare and financial services.

Mike is a Lean Six Sigma Black Belt and has designed numerous successful BPR programs. A Certified Coach. He is certified in numerous 360 assessment tools, such as the Leadership Circle Profile. In his spare time, Mike likes to do stand up, acting and improv and is fascinated about how this can be applied to the business world.

FINAL WORD FROM THE AUTHOR

Well, it is down to you now. If you were not aware of it before, 'quality' is a huge subject and has a presence in everything that we do, even the act of calling a plumber to fix a dripping tap is a quality issue. After reading this book, you will never see things the same again. A friend of mine once commented on this. He bought a wheelbarrow. The following day he took it back and told the salesman that it squeaked. The salesman said, 'That is OK, you just need to put some oil on its axle'. My friend replied indignantly, 'No, you oil it, it is your problem!' Maybe taking things a bit far but it is illustrative!

Best of luck.

Appendix
An account of the quality-related post-war history of Japan

This appendix gives some background information on the factual history of 'quality' in post-war Japan. There are many who are under the illusion that, mysteriously, an ageing American academic whose primary skill was the science of statistics applied to the analysis of data related to population census, somehow, in the course of a few lectures, transformed Japan from being totally devastated by the consequences of World War II so that in less than ten years the country was dominating world markets in almost everything from steelmaking, shipbuilding, cameras, white goods, brown goods and all things automotive. Some achievement, if it were true. Strangely, when he was discovered back in the USA some 30 years later, he was unable to produce the same magic in his own country, reference the so-called 'Rust Bucket States'. First of all, there was a cast of characters, almost all of which are conveniently ignored by the advocates of the esteemed professor and second, and most importantly, this deception is one of the primary reasons why the West has failed to match the performance of not only Japan but also China and the rest of South East Asia. It will continue to do so unless or until it confronts reality, gives up the myths and goes right back to first principles and starts from there. Hopefully, the following information taken from the internet might just help this process.

To look into this more deeply, I will mention some of the relevant names that were significant in that incredible revolution. I will mention just a few of the key people Americans first as in Japan's case, it was the impact of several organisations as much as it was individuals. This is not surprising when you know a thing or two about Japan because Japanese people are significantly more modest than their Western counterparts and have a natural tendency towards teamwork. Arrogance is just not accepted.

The key Americans include almost in order of appearance, General MacArthur, Charles Protzman, Homer Sarasohn, Dr Deming, Dr J.M. Juran, and Dr Feigenbaum. The Japanese main inputs came from JUSE, MITI, Zeibatsu (prior to and during World War II but disbanded by General MacArthur), other Japanese societies such as the Japanese Plant Maintenance Society, and individuals such as Ichiro Ishikawa, Professor Kaoru Ishikawa and Dr Junji Taguchi and several specialists from major exporting companies such as Toyota, Nissan, Canon, Mitsubishi and Panasonic, etc.

FACTS

After the Japanese surrender in 1945, General Douglas MacArthur was appointed supreme commander of American forces in Japan until the peace treaty in 1952. MacArthur was a powerful man with a clear vision of what he wanted to achieve. He had two main objectives. They were not necessarily stated as such but they were self-evident.

1. To change the culture of Japan as he saw it so that never again would they go to war again. At that point Japan had been at war for over 17 years!
2. To assist in the reconstruction of Japan so that it would not have to live on indefinite handouts from the USA.

To achieve the first objective, he did something rather dramatic. Effectively, he sacked all the top industrialists over the age of 45. Effectively, he did this by disbanding the all-powerful *zaibatsu*. This could be likened in some respects to the Confederation of British Industry in the UK.

Zaibatsu (Japanese: 'wealthy clique'), any of the large capitalist enterprises of Japan before World War II, were like cartels or trusts but were usually organised around a single family (see www.britannica.com/topic/zaibatsu). One *zaibatsu* might operate companies in nearly all-important areas of economic activity. The Mitsui combine, for example, owned or had large investments in companies engaged in banking, foreign trade, mining, insurance, textiles, sugar, food processing, machinery, and many other fields as well. All *zaibatsu* owned banks, which they used as a means for mobilising capital.

The four main *zaibatsu* were Mitsui, Mitsubishi, Sumitomo and Yasuda, but there were many smaller concerns as well. All of them developed after the Meiji Restoration (1868), at which time the government began encouraging economic growth. The *zaibatsu* had grown large before 1900, but their most rapid growth occurred in the twentieth century, particularly during World War I, when Japan's limited engagement in the war gave it great industrial and commercial advantages.

After the signing of the peace treaty in 1951, many companies began associating into what became known as enterprise groups (*kigyō shūdan*). Those created with companies that were formerly part of the big *zaibatsu* – Mitsubishi group, Mitsui group, and Sumitomo group (*qq.v.*) – were more loosely organised around leading companies or major banks; they differed most significantly from the old, centrally controlled *zaibatsu* in the informal manner that characterised each group's policy coordination and in the limited degree of financial interdependency between member companies.

The cooperative nature of these groups became a major factor in Japan's tremendous post-war economic growth, because, in the pooling of resources, the investments made by these groups in developing industries were large enough to make these industries competitive worldwide. This is never mentioned in Western accounts of the post-war history.

Also, as a consequence of MacArthur's dramatic action, it meant that immediately, the heads of these large organisations were gone, and that younger people were suddenly thrown into senior management roles but with no relevant experience. Under the old Shogun-style regimes, it would have been years before these younger people would have had the opportunity to take such positions and only after having been coached by their superiors to such a degree that they would have eventually slipped seamlessly into their shoes.

Now, these young people had nobody to copy so what is more likely than to attempt to copy the methods used by the victors in the war? There was a cultural element in deciding to do this and MacArthur knew it. So, he set up a major top-level intensive training programme. It lasted three weeks and was seven days a week and 12 hours a day. It began in 1945 and continued through to the end of the Occupation but even beyond that, the Japanese continued to run it themselves and it ran for a total of 23 years. The course was conducted by three key people who were all employees of the Western Electric Company in the USA. The leader was Homer Sarasohn and he was aided by his colleagues.

In four years this democrat in dictator's clothing (Homer Sarasohn) may have accomplished more than any economic dictator in history.

(Wood 1989)

Sarasohn teamed up with Protzman in 1948 to design and teach intensive management training seminars, the Civil Communication Section (CCS) Management Seminar ... After World War II, Japan's 'captains of industry' fortified Sarasohn's management values with Japanese Bushido values ... they ... produced the postwar 'miracle'.

(Yoshio Kondo, Baruch College, City of New York)

A leader's main obligation is to secure the faith and respect of those under him.
(Charles Protzman and Homer Sarasohn, CCS manual, GHQ Tokyo, 1950)

I gave over 500 lectures in every part of Japan from Hokkaido in the north to Kyushu in the south. ... CCS was the light that illuminated everything ... By the end of the (1950) CCS seminar we all knew we would catch up with the Americans.
(Bunzaemon Inoue, co-chair, CCS Seminars. Chairman, Sumitomo Rubber Industries)

The following is an extract from Ken Hopper's The Puritan Gift *(2007):*

General MacArthur's Civil Communications Section was very important for both Deming and Japan's success. Our ignorance of it is surprising. CCS drilled into Japanese top executives that they were responsible for good management. When one missed a CCS session, CCS was on the phone to remonstrate. 'Our slightest wish was their command', Protzman remembered(1) Deming would be saddened at CCS neglect. I know. When illness struck, he sent me a substantial sum to help me tell the CCS story. When Human Resource Management published my *Creating Japan's New Industrial Management*, Deming wrote to me 'This is just what I need'.

Late in 1969 Peter Drucker phoned to say he had someone I should meet. I asked no questions and turned up to find a near mythical Japanese Sensei (teacher), Takeo Kato. Kato started to tell me about CCS when the tall Polkinghorn appeared and Kato said, 'No man has done more for Japanese industry'. I was hooked. In 1979, my wife and I were treated like royalty in Japan at the invitation of Sumitomo Electric, Sumitomo Rubber and Matsushita Electric. Because we knew the great CCS engineers, dinners, waiting limousines, guides and interpreters were everywhere. The story of CCS is now well known in industry but has not reached business schools who prefer to see Japan's success as an inexplicable miracle. Simplistic shock treatment was given to Russia when the lesson from Japan was rebuilding requires work in depth including improving how factories run. The most let down were non-Asian developing nations.

Communications in Japan in 1945 were a disaster. The War Department approached US industry for help and able people responded including a young Homer Sarasohn who had impressed the US Army at MIT's Rad Lab, Charles Protzman, a 6 ft 4 in manufacturing superintendent with decades of experience from Western Electric and Frank Polkinghorn, a high engineer in Bell Labs. I came to know all well. Their ability and domain knowledge made it possible for Japan's unimpressive electrical manufacturers to become its world stunning consumer electronics industry. Influenced by Morgenthau, the US had shackled MacArthur with Secret Order JCS1380/15 to take no responsibility for the Japanese economy. CCS proposed that it combine seminars with working with its manufacturers to help them compete in world markets. Sarasohn loved to recount his 1949 confrontation before MacArthur with the large Economic and Scientific Section who argued the US would be giving away too much. MacArthur sat expressionless through both presentations, got up and walked to the door. Sarasohn thought, 'I've blown it'. MacArthur turned, pointed the stem of his corncob pipe at him and said, 'Go do it'. The rest is history. Japan's electronics industry would have a major influence on management in the rest of Japanese industry and the world.

Japan's specialists wanted a visit from Shewhart. Sarasohn refused until 1950. When he phoned, Shewhart declined for reasons of ill health. Sarasohn confirmed Deming's ability. At his request, ESS issued a formal invitation. Many have confirmed the importance of CCS. My brother Will has a selection at www.puritangift.com including an abbreviated chapter on CCS and Japan's extraordinary executives from our book *The Puritan Gift*, a *Financial Times* Top Ten Business Book of 2007.

The following is a reproduction of a letter sent to Sarasohn by the chief of the CSS showing his appreciation.

Dear Mr Sarasohn,

On the occasion of your departure from Japan, I wish to express my appreciation for the invaluable contribution you have made to the Occupation in your capacity as Industrial Engineer in the Research and Development Division of the Civil Communications Section, General Headquarters, Supreme Commander for the Allied Powers.

Your outstanding work in connection with the rehabilitation and re-orientation of the Japanese communications equipment manufacturing industry has materially aided in the task of re-establishing Japan as a completely stable nation.

Through the introduction of the ideas of scientific industrial management you have helped to raise the engineering standards and promote the use of modern manufacturing methods and practices to such an extent that the industry is now one of the most reliable and important adjuncts of the Japanese economy.

I wish to commend you for your outstanding work in introducing in Japan the modern concepts and practices of statistical quality control and for your development of the series of books in modern industrial management methods through which you have so ably indoctrinated Japanese management personnel in up-to-date techniques of production management, manufacturing methods, production engineering and design engineering practices.

Your advice and assistance on the many problems in the field of research and development in

the Japanese communications equipment manufacturing industry have contributed materially to the accomplishment of the Civil Communications Section's mission.

Please accept my best wishes for continued success in connection with your future endeavors.

Sincerely,
GEORGE I. BACK
Brigadier General, USA
Chief, Civil Communications Section

(*Homer Sarasohn and American Involvement... (PDF Download Available)*. Available from: www.researchgate.net/publication/46537675_Homer_Sarasohn_and_American_Involvement_in_the_Evolution_of_Quality_Management_in_Japan_1945-1950 [accessed 3 June 2018])

SUMMARY

Dr Deming's contribution was important, and this is undeniable, but with the passage of time the reality is quite different from the beliefs of many modern quality professionals. This is partly because it suits their personal interests to elevate his importance because it is a good story and if they are also consultants, they get reflected glory and partly because many of them have not taken the trouble to check facts.

References

N. I. Fisher (2008). Homer Sarasohn and American involvement in the evolution of quality management in Japan, 1945–1950. *International Statistical Review*.

K. Kobayashi (1985). Chairman NEC Corporation, 1985 address to Bell Laboratories.

Robert Wood (1989). A lesson learned, and a lesson forgotten, FORBES, Feb 6, 1989.

Bibliography

Mostly this is recommended additional reading to reinforce the text and to go into greater depth on the various topics. I have not specifically quoted anyone in particular in the text but have mentioned many, all of whom are relevant to the text where they are mentioned. All of them have a great deal to say and it is recommended that it is regarded to be a recommended reading list. Much relevant content can be found by googling the names and searching on YouTube.

Recommended reading in the order they are mentioned in the book:

Frederick W. Taylor – *The Principles of Scientific Management*.
Professor Kaoru Ishikawa – *What is Total Quality Control?*
Dr J. M. Juran – *Managerial Breakthrough and Others*.
AQAPs 1969 – Revised from Mil Q 9858.
ISO 9000: 2015.
Karl Marx (1869) – *Das Kapital*.
Adam Smith – *The Wealth of Nations*.
Frank and Lillian Gilbreth – The Management Theory of Frank and Lillian Gilbreth, Business.com.
Henry Ford – Various.
Phil Crosby – *Quality is Free* et al.
Zaibatsu – Various internet references.
Homer Sarasohn – Many internet references – Japanese post-war history.
Charles Protzman – Various references associated with Homer Sarasohn.
Dr Frank Polkinhorn – Also associated with Sarasohn and the SCAP work of General MacArthur.
Dr Edwards Deming – *Out of the Crisis*.
Dr A. Feigenbaum – *Total Quality Control*.
C. Drury and J. Fox: *The Reliability of the Human Inspector*.
D. Harris and F. Cheney – *Human Factors in Quality Assurance*.
Dr W. Shewhart (1931). *Economic Control of Quality of Manufactured Product*.
William Ouchi – *Theory Z Management*.
Bruce Tuckman – Forming Storming and Performing Theory – numerous references on the internet.
Dr Meredith J. Belbin – *Management Teams Why They Succeed or Fail* – many references.
Douglas McGregor – Theory X and Theory Y – many references on the internet.
Dr Genichi Taguchi – Many references on the internet. His main contributions are the quadratic loss function and his orthogonal arrays method for design of experiments.

Index